The *Routledge Environmental Humanities* series is an original and inspiring venture recognising that today's world agricultural and water crises, ocean pollution and resource depletion, global warming from greenhouse gases, urban sprawl, overpopulation, food insecurity, and environmental justice are all *crises of culture*.

The reality of understanding and finding adaptive solutions to our present and future environmental challenges has shifted the epicenter of environmental studies away from an exclusively scientific and technological framework to one that depends on the human-focused disciplines and ideas of the humanities and allied social sciences.

We thus welcome book proposals from all humanities and social sciences disciplines for an inclusive and interdisciplinary series. We favour manuscripts aimed at an international readership and written in a lively and accessible style. The readership comprises scholars and students from the humanities and social sciences and thoughtful readers concerned about the human dimensions of environmental change.

For more information about this series, please visit: www.routledge.com/Routledge-Environmental-Humanities/book-series/REH

Towards an Ecocritical Theatre

Playing the Anthropocene

Mohebat Ahmadi

Routledge
Taylor & Francis Group
LONDON AND NEW YORK

earthscan
from Routledge

First published 2022
by Routledge
4 Park Square, Milton Park, Abingdon, Oxon OX14 4RN

and by Routledge
605 Third Avenue, New York, NY 10158

Routledge is an imprint of the Taylor & Francis Group, an informa business

British Library Cataloguing-in-Publication Data
A catalogue record for this book is available from the British Library

Library of Congress Cataloging-in-Publication Data
A catalog record for this book has been requested

ISBN: 978-0-367-50088-7 (hbk)
ISBN: 978-0-367-50091-7 (pbk)
ISBN: 978-1-003-04874-9 (ebk)

DOI: 10.4324/9781003048749

Typeset in Bembo
by Apex CoVantage, LLC

For Melody, my lovely daughter

Contents

Figures

Acknowledgements

Working on this book has been such a privilege. I would be remiss if I did not acknowledge how much of an assemblage this book has been and how without many human and nonhuman actants it would never have been possible to finish this process.

I am grateful for the opportunity to thank some of the people who have shaped my thinking and made this book possible. The project began as my dissertation, and the research groundwork for this book took place at the University of Melbourne in the School of Culture and Communication. I am indebted to many professors and friends who have read parts of the work, engaged in conversations about its ideas, and otherwise supported this scholarship. Acknowledgements are due especially to the professors on my dissertation committee for their critical insights and generous guidance during this project's early stages. My sincere thanks go to my primary supervisor, Denise Varney, who provided me with invaluable and thoughtful dramaturgical feedback and the greatest assistance and support possible. Her steady flow of constructive criticism was of vital importance for teaching me how to be an academic and enabling me to sharpen my arguments and focus. I am also grateful to Peter Eckersall, Clara Tuite, Sarah Balkin, Peta Tait, Kate Rigby, and Thomas Bristow for providing some much-needed critical foundations and helpful suggestions at the very beginning, when this project was in the form of a PhD dissertation.

I would like to extend my special thanks to the playwright Chantal Bilodeau for her valuable perspectives on my ecocritical analysis of her play in Chapter 4 and for providing a platform for ecological thinking in theatre and performance studies through her blog "Climate Change Theatre Action." Sincere thanks to Jordan Hall and Miwa Matreyek for their generosity throughout writing Chapter 5 of this book. I am so appreciative of Jordan Hall for offering thoughtful and detailed comments and corrections and for granting me permission to include excerpts from her unpublished theatrical script in this book. I am grateful to Miwa Matreyek for her generous feedback and allowing me to use some images from her performance pieces. I would also like to thank E.M. Lewis and all the artists, photographers, and managing and artistic directors who granted me permission to use images of the plays and performances discussed in this book.

My thanks to the anonymous reviewers for providing valuable commentary, critiques, and feedback on the book proposal. This book would not have been possible without the guidance and support of its publisher. I would like to thank Routledge and, in particular, Rebecca Brennan, Grace Harrison, Rosie Anderson, and Matthew Shobbrook of Routledge's Environment and Sustainability list for all their help and encouragement in moving this book to the final stage of production. I would also like to thank the editors of the Environmental Humanities series, Scott Slovic, Joni Adamson, and Yuki Masami. In particular, I am extremely indebted to my eco-mother Joni Adamson who has greatly influenced the direction of my studies in the field of ecocriticism. Since I met her at the 2010 Environmental Change – Cultural Change Conference at the University of Bath (UK), her intelligence and encouragement have steered me in the right direction and helped me find my footing as I developed this project. Heartfelt thanks to her for her formative and generous support.

This book has also been inspired by the mentorship of other brilliant ecocritics in the Association for the Study of Literature and Environment (ASLE). Special thanks to Greg Garrard, Serpil Oppermann, and Serenella Iovino, as well as William Stroup, ASLE's mentor, for their scholarly direction and critical insights during the early stages of this project.

I am grateful to Michelle Niemann for editing and proofreading this book and for her thoughtful, constructive, and inspiring comments and suggestions. My thanks also to Catriona Sandilands and Ian Garrett for their generosity and their belief in the value of this work. I wish to express my gratitude to all my colleagues and friends in Melbourne, Toronto, and Iran who shared their time with me during the writing of this book. Enduring thanks to Parisa Shiran, whose warm presence and encouragement brightened the dark corners of this journey.

This book, and the adventures leading up to it, would not have been possible without the support of my family. Endless thanks go to my mother, the eternal flame of my life, for her unceasing love, support, and continuous encouragement in my quest for learning. I am deeply grateful to my brother and my sister for their encouragement and sympathy, and for uplifting my spirits when I was down. I am equally indebted to my father-in-law and mother-in-law for their understanding and boundless support every step of the way. Last, but not least, I express my gratitude to my husband, Khalil, who throughout this entire process has always been a source of love, grace, compassion, encouragement, and unwavering support.

1 Where ecocriticism meets theatre in the era of the Anthropocene

Despite the importance of reconsidering theatre in the framework of Anthropocene events, dramatic works about global warming and mass extinction have been scarce. According to Julie Hudson, in spite of the apparent dramatic appeal of the climate-change debate, theatre has been distinctly silent on the subject of anthropogenic climate change and its implications for the social, psychic, and cultural life of humankind (260). Hudson contends that, when Kristen Shepherd-Barr listed "no fewer than 82 science plays written between 1992 . . . and 2004 without directly discussing climate-change science" in the Appendix to her *Science on the Stage* in 2006, "the presence of climate change and its science on the stage was barely visible" (260). Furthermore, in "Staging Climate Change: The Last Ten Years" (2014), Bradon Smith investigates theatre's delayed creative process in discussing climate change. However, as he argues, "the all-encompassing nature of the implications of climate and environmental change" makes it "a compelling subject for drama" (n. pag.) while theatre also has the potential to suggest innovative ways of thinking about the reality of climate change.

Why have representations of the global ecological crisis been rare in theatre? The urgency of the issue prompted the Ashden Directory of Environment and Performance established by Wallace Heim to convene a round table in 2005 to discuss the question, "Why are there no plays about climate change?" (Smith, "Climate Change in Performance" 9). There are many possible answers for such questions. In his 2008 blog post "Why Theatres Don't Touch Climate Change" on Ashdenizen, Robert Butler writes that it is hard to see how the spatial and temporal impact of individual actions can be represented within the conventional patterns of cause and effect in dramaturgy. He argues that theatres resist engaging with climate change because they "think climate change is about science," rather than about "drama's core themes: human relationships, the way we live, what we value" (Heim and Margolies 166). The theatre scholars Wendy Arons and Theresa May also expand on a key challenge playwrights face in telling compelling stories about disasters such as Hurricane Katrina and volcanic eruptions, as well as about glaciers, rivers, and species. According to these scholars, such stories go beyond the human scale of time and space, and "so even when a playwright strives to foreground ecological issues on stage, the

DOI: 10.4324/9781003048749-1

stories are hard to contain" (*Performance and Ecology* 4). While materialist theoretical approaches to the nonhuman world are effective ways to think about how theatre engages with global ecological changes, they face challenges of complexity and scale given Anthropocenic alterations. In *Slow Violence and the Environmentalism of the Poor* (2011), Rob Nixon characterises global ecological changes of environmental toxicity, climate change, and extinction as "long-dyings" and slow forms of violence that occur "gradually and out of sight" and are "dispersed across time and space" (2). While we live in a world that is attuned to immediate and spectacular experiences, Nixon wonders "how can we convert into image and narrative the disasters that are slow moving and long in the making. . . . How can we turn the long emergencies of slow violence into stories dramatic enough to rouse public sentiment and warrant political intervention?" (3). These key questions in relation to the challenges of representational strategies are in sync with concerns about the barriers that theatre and performance encounter in representing the complexity of slow-moving ecological disasters.

To explain why climate change is not amenable to traditional modes of representation in art and theatre, I refer to eco-theatre scholar and pioneer Una Chaudhuri's recent collaborative "Ecocide Project" (2014) co-authored by Shonni Enelow. In it they argue, first, that theatre deals with more concrete and comprehensible topics. These include narratives of family and domestic issues and representations of sensational and dramatic forms of terror, rather than abstract global threats such as climate change (23–4). Second, dramatising climate change, with its "particular temporalities," in theatrical space poses an intellectual challenge in terms of both what it means to be human and knowledge production in the humanities and social sciences. The problems caused by the phenomena of climate change are, in particular, related to the "*time scales*" and "*conceptualizations of human agency*" (25). The final barrier to climate-change representation, for Chaudhuri and Enelow, is the fact that climate change is "the aggregate result of countless dispersed behaviours and practices" (27). In other words, the "Anthropo" in "Anthropocene" is not singular but "a collective that is in some fundamental and scary way in excess of each of us and all of us, a collective that acts outside our control" (27). This point calls for a non-anthropocentric approach that presents a new version of humankind in entanglements with other humans and nonhumans. In spite of such conflicts between the urgency of climate change and the aesthetic challenges it offers, I argue that a critical understanding of theatre as a broad and collaborative field of art – with the capacity to embody and perform imaginative, immediate, and affective possibilities and to present new epistemological and ontological frameworks of thinking, feeling, and viewing our place in the world – articulates an ever-evolving connection between art and science in the context of a changing environment.

Proposed in 2000 by Nobel prize-winning atmospheric chemist Paul Crutzen and ecologist Eugene F. Stoermer, the term Anthropocene designates a proposed new geological age that follows the Holocene. This new geological age, characterised by dramatic and extensive human-induced changes in the Earth's biosphere, underlines the pervasive, decisive, and "central role of mankind in geology and ecology," the impacts of which "will continue over long periods" (Crutzen and Stoermer 17). Such impacts include anthropogenic carbon emissions, widespread levels of pollution, altered ocean chemistry, and changes in global temperature and climate along with increasing extinctions of plant and animal species. The complex challenges and perceptual shifts that arise from this new version of humankind whose ecological footprints are inadvertently left on all human and nonhuman bodies in the biosphere prompt a paradigm shift in the arts and humanities, requiring scholars to radically rethink common conceptions of human subjectivity and the relationship between the human and the nonhuman. *Towards an Ecocritical Theatre: Playing the Anthropocene* aims to make a timely intervention in theatre studies by addressing theatre's humanist bias and extending existing scholarship that has situated theatre within an overarching narrative of the Anthropocene. The book engages with major new approaches in environmental humanities and arises out of the need to address how Anthropocene thinking engages different modes of theatrical representation, as well as how the theatrical apparatus is capable of embracing and rising to the representational challenges of the current global ecological crisis and changing interactions between humans and the nonhuman world. Although considering environmental issues in performative terms can be challenging, the book celebrates this intersection and models imaginative and informative ways of communicating about our environmental future.

This book identifies important drama that has not been fully analysed. As a body of plays and performances, the works I examine reflect how theatre is informed by, and informs, problems of scale and agency in an era of human-induced alterations to the environment. In exploring these points of tension, the book is grounded in a close and in-depth reading of international Anglophone plays and performances by Caryl Churchill, Stephen Sewell, Andrew Bovell, E.M. Lewis, Chantal Bilodeau, Jordan Hall, and Miwa Matreyek, who have taken significant steps towards re-orienting theatre from a human-centred focus to an Anthropocenic focus within the emerging field of ecocritical theatre. I categorise the works as examples of ecocritical theatre that I define as a new approach reflecting on how theatre – as the most anthropocentric form of art – might represent the nonhuman world by challenging the anthropocentrism that underpins the Anthropocene; dramatising the social, cultural, political, and economic dimensions of the environmental crisis; and integrating with new materialist and eco-global thinking. I show that the plays and performances are themselves ecocritical – rather than just ecological or environmental – as they embody the interactions between the performing arts and science, between human cultural expression and environmental concerns, allowing an interdisciplinary consideration of new material and multispecies agencies that involve the problem of scale that the Anthropocene entails. In doing so, this book draws on critical theories to engage in textual and performance analysis of the selected theatrical works. While the context of performance is addressed, this study primarily focuses on the close reading of plays and dramatic literature rather than on performance theory – except for Matreyek's performances in Chapter 5. As my project connects to literature and cultural theory, it has potential to fill an important gap in the literature and theatre studies and reach readers widely across the humanities.

While the number of plays and performances that examine Anthropocenic concerns such as climate change has been growing around the world and in different cultures, I primarily focus on a broad spectrum of ecocritically oriented theatrical works written and performed in Anglo-centric contexts in the United Kingdom, the United States, Canada, and Australia. The works were written during the late twentieth and the twenty-first centuries: precarious times of escalating conditions for global ecological crises. Western countries' dominant contribution to global emissions and, consequently, to the global climate crisis and mass extinction informs my choice of national contexts.[1] Critiques of Western countries call for close examination of their theatrical production in order to get a new cultural and ecological conception of human-nonhuman relationship in the context of a rapidly changing world. Therefore, I have selected plays and performances from Western, Anglophone countries that shift the focus from the anthropocentric perspective that has conventionally dominated Western theatre and performance to one that defines humans in mutual interaction with the nonhuman world. This shift also gives new insights into cultural and political ideologies that form the basis of some of the most important environmental concerns of the past and present (Richmond 123). Hence, although these theatrical works respond to information, images, and discourses that are circulating in English and respond to environmental change in the context of particular Anglophone cultures, my principle aim is to lay the groundwork for discerning a genre of Western plays and performances that posit a global and inclusive perspective through a radical rethinking of some of the principle assumptions made about human-nonhuman relationship. I will discuss the environmental contexts particular to each set of works in more specific terms and provide focused contextual discussion in each chapter.

Accordingly, *Towards an Ecocritical Theatre* approaches the concept of the Anthropocene not only in terms of its stratigraphic meaning but also as a new mode of narrative about the present state of humanity and its environment. My use of this concept is also an attempt to open new discussions about ecocriticism – as one of the strongest contributing subfields of the broad multidisciplinary approach of "the environmental humanities" and, in particular, new cultural and ecological insights within the field of theatre studies. This book takes up the ecocritical concerns of the Anthropocene context to reexamine how contemporary theatre, through its liveness, immediacy, flexibility, performativity, physicality, and communality, participates in such new acts of storytelling. Following the premise that Anthropocene events such as pollution, climate change, and mass extinction are always ecologically, politically, and culturally entangled with any number of humans, nonhumans, and materials, *Towards an Ecocritical Theatre* examines the following questions: How do works of theatre and dramaturgical strategies shed light on these entanglements? How can the theoretical concerns of ecocriticism provide a platform for such an examination? How does theatre define and expand the sense of community and collective imagination, ethics, responsibility, empathy, and action needed to deal with our current crisis? How will reshaping and rethinking values and storytelling towards new theatrical possibilities – without losing sight of the trail we have left behind – make us alert to devastating ecological events while stirring up potent responses and help us imagine alternative possibilities that might yet materialise?

While the following sections show that the ecocritical turn is an exciting new territory for theatre studies, they also examine important barriers to ecocritical theatre. However, rather than focusing on these limitations, this book aims to contribute to assessing how theatre embraces the challenges of human engagement with the nonhuman world. Further, because questions of scale and agency are the main challenges identified in both ecocriticism and humanistic discourse on the Anthropocene, combining insights from these two fields provides an efficient theoretical approach to this study. Hence, a discussion of contemporary trends in ecocriticism – building on the theoretical approaches of eco-cosmopolitanism, new materialism, and object-oriented ontology – and their agential and spatial relation to the idea of the Anthropocene gives new insight into how theatre yields a novel understanding of the emergent field of the environmental humanities. In that way, a theatrical work allows for a pivotal mode of ecological communication that can reveal, enact, and indeed shape creatively, thematically, and performatively the co-evolutionary interplay of human activity and the environment, understood in its broadest sense.

Locating theatre in the Anthropocene: the agential configuration of scale framing

While there has been quite a delay in the theatrical response to the real urgency of climate change, a significant body of early examples came into focus in the UK with Clare Pollard's *The Weather* (2004), Churchill's 2006 climate-change libretto for the London Proms, *We Turned on the Light*, and the 2007 *Crown Prince* by John Godber (Hudson). Following these early examples, the number of climate change theatre works has steadily increased. In 2016, Shepherd-Barr said that "only since around 2008 have plays been produced that deal with it [climate change] overtly" ("An Interview"). For instance, the interest in climate change in British theatre flourished in the following works, which also deal with themes of extinction and other ecological catastrophes: Steve Waters' *The Contingency Plan* (2009), Mike Bartlett's *Earthquakes in London* (2010), and *The Heretic* by Richard Bean (2011).[2] There has also been a steady flow of excellent theatrical works about anthropogenic climate change in other parts of the world. For example, Bilodeau, in her blog "Artists and Climate Change," compiles plays in which climate change is featured either as a central or peripheral issue; they are mainly from the USA, the UK, Canada, and Australia, along with a few works from Ireland, Scotland, Sweden, Spain, Germany, Brazil, South Africa, and Bulgaria. Furthermore, Bilodeau's 2015 project, the international Climate Change Theatre Action (CCTA) – a collaboration by The Arctic Cycle, the Centre for Sustainable Practices in the Arts, NoPassport Theatre Alliance, and Theatre Without Borders – consisted of a worldwide series of readings and performances of climate-change plays intended to raise awareness about and support discussion around the United Nations Conference of the Parties (COP) meetings, the decision-making body of UN Climate Change. As a network for connecting playwrights, the CCTA insists on building ecological knowledge in theatrical response to climatic disequilibrium, calling for a reconfiguring of theatre and its theories. CCTA is a global biennial event that has been held in 2017 and 2019, with the upcoming iteration scheduled for Fall 2021. Further, the National Arts Centre's English Theatre Cycle on Climate Change, co-curated by Bilodeau in June 2020, engages a broad theatrical community of academics, artists, and scientists to provide a multidisciplinary forum for sharing ideas around the question of how the performing arts can respond to climate change.

The Anthropocene age and the threat posed by climate change, Chaudhuri contends, "provide the contemporary conditions for a new sub-genre of theatre practice" ("Research Theatre" 2). This sub-genre is "an up-dated ecotheatre, dedicated to putting the vast resources of live, embodied performance at the service of the program of radical reimagination called for by the perilous predicament we find our species – and others – in today" (2). This updated form of ecological theatre, which is built upon a reconceptualisation of the very idea of human agency, does not follow the methods of modern drama, which centralises humans and human thought. In place of traditional forms of drama such as the plays of Ibsen, Chekhov, Pirandello, and Pinter, or what Chaudhuri describes as "the drama of big ideas brought 'home', of philosophy domesticated and thinking *located*," she proposes new dramatic strategies, which she terms "the drama of bad ideas" ("Anthropo-Scenes" 304). This newly invented dramatic form, as she puts it, is indebted to Martin Puchner's work, *The Drama of Ideas* (2010), which "traces the history of a philosophical modern drama that variously plays out the proposal, lodged in Plato's dialogues, for a theatre of embodied and situated ideas" and "theatre as a philosophical testing ground for the mutually shaping encounter of bodies, individuals and situations" (314). Dramas of bad ideas, however, "develop their ecological perspectives by deliberately distorting that confident tradition, producing an abject inversion wherein human beings and their social structures are revealed to be a breeding ground for ecological mistakes" (314). These forms of drama stage humans against huge forces, such as climate change and biodiversity loss, and the crisis of scale born of such phenomena. I am inspired by Chaudhuri's thinking on this novel sub-genre of theatre practice, mindful of how theatrical imagination, protocols, and production respond to the unprecedented and incommensurable scale and agency of dramatic ecological changes. I build upon her innovative approach to show how the plays and performances I analyse here are new forms of theatre that complicate and distort the conventions of the drama of ideas by being located in the wider temporal and spatial framework of the Anthropocene. The ideas these works embody and dramatise are problematic reflections of our current ecological disruptions, enabling reconfigurations and reformulations of the human story across many scales.

Chaudhuri also introduces the concept of "fifth wall dramaturgy." For Chaudhuri, this shorthand term characterises "new strategies that will align the theatrical apparatus – its conventions, protocols, and possibilities – with the altered conceptual frameworks offered – or necessitated – by climate change" ("Fifth Wall" n. pag.). In addition to "reaching across the fourth wall" (n. pag.) – that is, eliminating the conventional distance between actors and spectators – as one way to engage with the ecological crisis, Chaudhuri calls for the breaking of a "fifth wall" – that of anthropocentrism. For Chaudhuri, stretching across a fifth wall makes the dramatic frame not only look beyond the national and social boundaries that twentieth-century playwrights focused on, but also "expands it temporally, pushing past cultural history to locate the human story in the deep time of the earth" (n. pag.). Because the planet has undergone radical transformations, which were induced by former and recent anthropogenic activities but are now far beyond human power, we must retell the human story as part of a deep-time narrative of the earth. Reconfigured by geological time scales, Chaudhuri writes, the protagonist of fifth wall dramaturgy is a complicated human described as the "evolving phenomenon," who is compelled to try out "forms of ecological consciousness that are more attuned to the scale and complexity of climate change" (n. pag.). However, to make the ecological consciousness derived from this new theatrical experience real and intimate, the expansion of the dramatic frame should also preserve "the voice, body and emotion" (Chaudhuri, "Fifth Wall" n. pag.). This ecological view of theatre depicts a mesh of interconnection and coexistence that breaks the fifth wall of anthropocentrism. Using these frameworks in reading the theatrical works in this book enables me to show how an engagement with the problem of scale disrupts the humanist bias of theatre, provoking new modes of theatrical inquiry that envision a scale beyond the human and realign our ecological culture, art, and intimacy with geological time. Moreover, *Towards an Ecocritical Theatre* examines such scalar shifts by focusing on the problem of agency in order to give expression to the stories inherent in nonhuman actants.

Living in the Anthropocene requires reconsidering cause-and-effect relationships because the complexity of climate change – together with the agency of all kinds of energy and matter – exceeds the goals and intentionality of the humans who are causing it. In other words, according to Ursula K. Heise, the Anthropocene, "at first sight, seems to rely on too simple an understanding of the human and too anthropocentric an approach to historical agency" (*Imagining Extinction* 233). However, she adds that if the Anthropocene is understood "as a shorthand for the self-reflexive moment in which humans confront the pervasive intentional and unintentional transformations of planetary ecology and geochemistry they have brought about, it can serve both to complicate what we usually mean by 'human' and to offer opportunities for redefining it" (233–4). Seen through this perspective, theatre might be seen as part of the "self-reflexive moment" that Heise talks about, serving as a material expression of the fact that human agency is part of a planetary situation, of geophysical and ecological systems acting on a scale beyond human, and is defined by its material dynamics and interactions.

To develop this different sense of agency in "our telling of the human story" requires a development of "multiple-track narratives" (Chakrabarty, "Brute Force" n. pag.). The massive scale of human-driven phenomena such as climate change and mass extinction that mark the geological age are far beyond the scope of our imagination as they cannot be viewed through a single lens. Seen in this light, knowledge production also confronts new problems of scale, since the Anthropocene becomes ungraspable given unprecedented and widespread ecological change. In the article "Agency at the Time of the Anthropocene" (2014), Bruno Latour underlines the significance of knowing how to "tell our common geostory" (13). Telling our geostory, or the narrative of the entanglement of humans and nonhuman agencies, requires that we reform our knowledge about and imagination of subject and object. In order to open a different path for telling our story in the Anthropocene, we need to revise the idea of agency by addressing all the threads and all the agents of geostory, including living beings and nonliving forces, because "[t]he point of living in the epoch of the Anthropocene is that all agents share the same shape-changing destiny" (15). According to Latour, this destiny cannot be represented through the older modernist distinction between subjectivity and objectivity: "Far from trying to 'reconcile' or 'combine' nature and society, the task . . . is on the contrary to *distribute* agency as far and in as *differentiated* a way as possible" (15). Given the complex and dynamic ecological systems, the interactions between different agents and the emergent effects that arise from such unexpected and unimaginable dynamics need to be assessed across a wide range of scales. For example, we can approach the decentralised distribution of the agency of climate change across quantum scales, individual human bodies, local scales, and planetary scales. This complex perspective on the world as always moving and changing, rather than remaining static and consistent with linear conceptions of cause and effect, asks for new modes of imagination and storytelling that enable us to think of the material world as engaged with the agential configuration of scale framing. By closely considering theatre through blurred boundaries, dynamic interactions, and diffuse agencies across various sales and analysing these conceptual problems through theatre, this book critically reflects on the ties that bind them together. In doing so, I show that theatre represents global ecological crisis differently and across manifold scales, enacts the entanglement of human bodies with strange nonhuman forces, and disrupts the humanist bias of theatre.

The distribution and differentiation of agency as well as the question of scale – taken up by theorists Jane Bennett, Karen Barad, Donna J. Haraway, and Stacy Alaimo in the field of new materialism, Timothy Morton in object-oriented ontology, and other scholars such as Timothy Clark and Heise – are the key challenges of the Anthropocene that this book examines. Hence, my version of the Anthropocene – along with its cognitive, perceptual, cultural, and political challenges[3] – focuses on a reconceptualisation of the problems of scale and agency and a paradigm shift from human-centric worldviews to corporeal and collective planetary senses. Drawing on conceptual frameworks informed by these thinkers, I show how the work of contemporary eco-artists and playwrights indexes ecological thinking through interrogating the agential and spatial dimensions of the Anthropocene, while questioning the anthropocentric status of theatre. However, while some plays in this book raise ecological awareness and involve environmental problems without centering them, the other innovative recent works explicitly undertake and thematise ecological projects and show a greater perceptiveness of the scale, agency, and complexity of global ecological changes.

The plays considered in Chapters 2 and 3 are read in association with a newly realised planetary context. This realm of possibility in rereading the plays problematises the tendency to privilege human-centred cultural paradigms for theatre and performance, moving ecological realities from the background to the foreground. For example, Sewell's domestic scenario and Bovell's family saga in Chapter 3 show how the meaning of our reality, selfhood, everyday life, art, and psychology now and in the future are materially affected by the complex phenomenon of climate change that seems so proportionately out of scale. The question of scale as a central challenge in thinking about the Anthropocene outlines the crux of my ecocritical readings of these Australian plays. Alternatively, Chapters 4 and 5 highlight the uses of theatre as a critical tool for reading and imagining the Anthropocene, employing forms that directly address its features. The plays and performances in these chapters provide new perspectives on the shift happening in theatre in light of dramatic ecological changes and deal with those changes in a direct and obvious manner. For example, Hall's *A Brief History of Human Extinction*, known as an Anthropocene play, provides a distinctive new voice by dramatising the aftermath of complicity in the global risk scenario of a mass extinction. An analysis of planetary imagery and collaborative space in this speculative performance demonstrates explicitly and innovatively humanity's troubled planetary relationship and a dynamic performance of material communications and meanings in the face of the earth's mass extinction crisis.

By bringing theatre in line with contemporary ecocritical debates (eco-cosmopolitan and material trends) and considering theatre's capacity for ecocritical thinking, this book arrives at novel ways of comprehending the use of theatre and performance for comprehending the Anthropocene. Further, while my ecocritical analysis produces new insights into each of the plays and performances, we also arrive at a unique understanding of ecocritical thinking in the Anthropocene through the works' contribution to the field and the way ecocriticism is encoded in a theatrical work. The works offer this contribution by dramatising a deeply entangled environment of divergent spatial and temporal scales and agencies and a non-anthropocentric view of the world that are the main ecocritical issues of the Anthropocene. Overall, each play in this study attempts to comprehend global ecological disasters by resisting the ontological rift between humans and nonhumans and relating ecological changes to stories of love, loss, and family while remapping the concepts of subjectivity, agency, scale, and ethics in the face of emerging ecological realities.

The ecocritical turn in theatre studies

In the article "Performance and Ecology: A Reader's Guide" (2005), eco-performance scholar Wallace Heim refers to Joseph W. Meeker's *The Comedy of Survival: In Search of an Environmental Ethics* (1974) as one of the founding works in the field of theatre ecocriticism. By interweaving dramatic forms of comedy and tragedy with scientific ecology and connecting literary and environmental studies, Meeker addresses a new narrative of human-nature relationship. This focus suggests that if theatre that reflects the complexities of living in a world affected by human-induced environmental changes is needed, and if theatre is to model new ways of thinking and communicating about our environmental future, then scholars must reflect on how theatre works as an active ecocritical medium. In a special issue on ecology in the US journal *Theater* in 1994, Erika Munk complains that "our playwrights' silence on the environment as a political issue and our critics' neglect of the ecological implications of theatrical form are rather astonishing" (5). Munk's urgent plea for a reinterpretation of dramatic forms and new scholarship about ecological theatre was reinforced almost simultaneously by Chaudhuri and Bonnie Marranca, though from different perspectives. While Marranca, in her 1996 book *Ecologies of Theatre: Essays at the Century Turning*, deploys ecology as a metaphor and a theoretical approach to performative analysis, Chaudhuri insists on attending to the material rather than metaphorical senses of ecology.

In fact, Chaudhuri, in her 1994 article, "'There Must Be a Lot of Fish in That Lake': Toward an Ecological Theater," was the first to approach performance discourse and dramaturgy via theories that emphasised the agency of the nonhuman world. For Chaudhuri, the common belief in the humanistic features of modern theatre and drama, particularly naturalism and realism, produces a "programmatically anti-ecological" stance in which ecology is treated as a metaphor and universalising concept (24). She identifies the first stages of "a remapping of humanism" (30), outlining "a new materialist-ecological theater practice that refuses the universalization and metaphorization of nature" (24). Chaudhuri stresses that the material engagement of theatre with ecological action "is not merely desirable in terms of the theater's own standards of social seriousness and political relevance: it is also crucial to the future of the ecological movement itself, and hence to the future as such" (25). She points out that the ecological crisis is a crisis of values, contending that "[e]cological victory will require a transvaluation so profound as to be nearly unimaginable at present. And in this the arts and humanities – including the theater – must play a role" (25). In the same year, Larry Fried and May published their book *Greening Up Our Houses: A Guide to a More Ecologically Sound Theatre*; in it, they offer sustainable material paradigms and practices for ecological design in all elements of the production-making process and scenographic methodologies. This publication prompted the emergence of ecoscenography and new approaches to recycling practices in all stages of scenographic thinking and theatre production.

After these beginnings, ecocriticism is only now finding solid ground in the field of theatre studies. It was over a decade later that May, in "Greening the Theater: Taking Ecocriticism from Page to Stage" (2005), alluded to both Munk's and Chaudhuri's calls for the development of an ecological theatre when she wondered "[w]hat accounts for theater's absence from ecocritical discourse, indeed from the environmental movement?" (84). In her 2010 article "Beyond the Nature/Culture Divide: Challenges from Ecocriticism and Evolutionary Biology for Theater Historiography," Arons notes that "only a handful of theatre scholars and historians to date have attempted . . . studies of the relationship of theatre to the environment" (149). In *Readings in Performance and Ecology* (2012), Arons and May explain how the two concepts of "performance" and "ecology" used in the title of their volume do not easily or readily "share space together, either materially or ontologically" (1). For the writers, this culturally inherited contradiction explains why "at the beginning of the second decade of the twenty-first century, ecology and environment are not only underrepresented and underthematised on the Western stage, but also undertheorized in theater and performance scholarship" (1). This situation may be related to the field's longstanding anthropocentrism and theatre's representation of exclusively human actions, perspectives, conflicts, communication, and cultural ideologies. To theorise and thematise the nonhuman world in theatre and performance, Arons and May argue that practitioners and scholars must reconceptualise preconceived binaries between culture and nature, humans and nonhumans (1–2).

Over the last 15 years, and especially in the last 5, such a new perspective in the field of theatre studies has been increasingly taken up by theatre scholars such as Alan Read, Baz Kershaw, Carl Lavery, Deirdre Heddon, and Downing Cless, among others. These scholars have contributed to the study of ecology and theatre, showing how theatrical form might accommodate new understandings of the human-nonhuman relationship, albeit in different ways and from particular perspectives.[4] The growth of the field has also included websites and free-web directories (Ashden Directory: Bringing Together Environmentalism and the Performing Arts established by Heim and HowlRound Theatre Commons). Furthermore, several major conferences, symposia, and international theatre festivals have been convened in this area and in response to the question of how performing art and drama respond to the question of sustainability, the intersection between humans and nonhumans, and warnings of climate change.[5] Together, these new initiatives in the field of performing arts practice and theatre studies scholarship address theatre's contribution to altering social and cultural values in response to current environmental challenges and announce a burgeoning movement in the field.

In the introduction to *Theatre Ecology: Environments and Performance Events* (2007), Kershaw addresses the seeming "ambivalence of theatre in the face of a calamity for humanity" (10). In light of the increasing interest in topics such as globalisation, environmentalism, and environmental risk in the third millennium, Kershaw insists on the importance of situating theatre within the context of the current ecological crisis, calling for substantial attention to be given to changing "the ecological qualities of the theatre phenomenon" (10). He asks: "What were the 'natures' of theatre and associated performances in times when influential voices increasingly predicted the 'end of nature'?" (10). This question evokes resistance to the traditional human-centred constitution of theatre in the face of our earth's environmental cataclysm and the end of nature. Further, in *Performance and Ecology: What Can Theatre Do?* (2018), Lavery writes that theatre's ecological contribution "resides in the more oblique possibilities inherent in the theatrical medium itself" (1). For Lavery, the theatrical medium – the art form that is supposedly obsessed with the anthropos – "is, in the very interplay of its internal mechanisms, always already 'more than human'" (3). In contemporary theory and practice, as Lavery argues, theatre might be regarded "as a form of ecological doing" rather than "a pre-existing script" or "a relatively stable, linguistic artefact":

> theatre is seen as a predominantly performative medium, that is to say, as something embodied, ephemeral and affective, with the result that the fundamental concern of scholars is no longer to decipher what the theatre text means but rather to focus on what the theatre medium 'does'; in how it might be figured as a form of ecocritical doing, and in how its dramaturgical distribution of human and nonhuman bodies in actual time and space creates sensations and experiences in the here and now.
>
> (2)

In sync with Chaudhuri, May, Kershaw, and Lavery, I integrate eco-global and new materialist thinking with theatre and performance, as this framework offers an alternative to exclusively anthropocentric systems of communication and foregrounds an awareness of the potential of a theatrical work as an ecocritical practice.

Because ecocritical theory can provide a platform for theatre's efficacy in fostering ecological cultures, I address the few works that have reflected on the ecological turn in theatre criticism. Chaudhuri's 1994 essay initiated a reexamination of canonical modern dramatic texts, including Henrik Ibsen's *The Wild Duck* and Anton Chekhov's *Uncle Vanya* and *Seagull*, and contemporary works such as Tony Kushner's *Angels in America* that feature nonhuman animals and landscape as dynamic forces. These plays are notable forerunners of environmental theatre due to their obvious concern with and critical observations about the clash between the human and the nonhuman worlds. In her articles

"Greening the Theater: Taking Ecocriticism from Page to Stage" (2005) and "Beyond Bambi: Toward a Dangerous Ecocriticism in Theatre Studies" (2007), May analyses canonical examples of the theatre that anticipate later understandings of the intersection of theatre and ecology. For example, May applies an ecocritical lens to Rodgers and Hammerstein's *Oklahoma!* and Arthur Miller's *Death of a Salesman* "to unmask the ways in which the arts have often been complicit with the causes of environmental degradation" ("Greening the Theatre" 90). May thus provides models for dramaturgical research on the history of drama's environmental impact. In his book *Ecology and Environment in European Drama* (2010), Cless takes a different approach, offering ecocritical interpretations of canonical examples of Western European drama that are widely considered anthropocentric and showing how these plays trace an awareness of the relationship between humans and the environment. His case studies are arranged chronologically from Aeschylus's *Prometheus Bound*, Euripides's *Bacchae*, and Aristophanes' *The Birds*, through Marlowe's *Doctor Faustus* and Shakespeare's *Midsummer Night's Dream* and *The Tempest*, to Ibsen's *An Enemy of the People* and Chekhov's *Cherry Orchard* and finally Giraudoux's *The Madwoman of Chaillot*. In his final chapter, Cless describes the encounter between climate and landscape in Brecht's *Mother Courage and Her Children*, "the de-natured void out the windows" in Beckett's *Endgame*, and the mysterious and barren landscape in *Waiting for Godot* (166). In reconsidering such plays, Cless highlights the ecological dynamics of canonical dramatic works written at a time when environmental concerns such as global warming were not explicitly part of public discourse. The analysis of theatrical works, and even the reinterpretation of the dramatic canon in conjunction with the contemporary ecological crisis, repositions environmental concerns and images in these works. Rather than being merely metaphors for the human condition, these concerns and images are reinterpreted as material embodiments of the world affected by the uncertainty of ecological change.

Theatre ecocritics have opened up opportunities for investigating ecological themes and communities in order to get insight into the potentialities of theatre in engaging with the urgency of current environmental issues. While ecocritical theories and concepts are running through theatre and performance studies, there is still a continued need for the scholarship to undertake ecocritical thinking and, in particular, its global and material perspectives on human-nonhuman relationships. I will, in the following sections, explore how key concepts in the eco-cosmopolitan and material trends of ecocriticism can be integrated into the analysis of dramatic literature and cultural theory. These theories' shared understandings of materiality give us new insight into how the plays and performances examined in this book respond to complex and disorienting sorts of interconnection between humans and nonhumans in the tangled mess of the Anthropocene. My selection of theatrical works also provides numerous examples of this kind of interconnection, shifting our perspective about subject and object into a wider scope. For example, my analysis of Churchill's plays in Chapter 2 explores the dark underside of global-scale ecological interconnectedness in *The Skriker*'s ecological allegory and *Far Away*'s worldwide ecological warfare by showing how ecocritical theatre can dramatise dynamic entanglements between human characters as material entities and a vital assemblage of nonhuman beings, things, and environmental forces. The Australian plays in Chapter 3 highlight this interconnection through the performative narrative of climate change, which acts as an expanded agential force across a field of human bodies and nonhuman objects and displaces the conventional human-centred form of drama. The ecologically themed theatrical works in Chapters 4 and 5 focus on the staging of stories and imagery that leads to a new understanding of ecological thinking, not only beyond human-nonhuman divisions but also across agential, spatial, temporal, and geological boundaries. These works contribute to shaping Anthropocenic forms of theatre and raising awareness about conceptualising life in and beyond the Anthropocene. Each play and performance constitutes a valuable site for demonstrating how theatre frames new ways of seeing and knowing the world, develops a non-anthropocentric ethical stance, and rethinks concepts and ideals that have, until now, been principal in our understandings of human and nonhuman scales and agencies. By opening a dialogue between science and theatre, *Towards an Ecocritical Theatre* questions and experiments with the tradition of anthropocentric representations of dramatic forms, prompting us to retell human stories from the perspective of nonhuman agencies in light of the Anthropocene.

Towards a global ecocritical turn in theatre studies: the challenge of scale

Early canonical examples of ecological theatre depict the gap between human social and economic concerns and the nonhuman world in local or regional contexts, such as via the discrete garden in *The Cherry Orchard*. In contrast, I deploy a more globalised framework in my ecocritical exploration of plays and performances in this book. This shift in perspective demands a redefinition of the human and the nonhuman. To answer Kershaw's question, "[i]n what ways has the theatre been unavoidably embroiled in the ecological mess that is climate change?" (10), new variants of ecocriticism are needed. In the face of ecological imperatives such as climate change that are planetary and cumulative in scale, such new trends demand that all scales be examined, from the immediate, embodied, and personal footprints of ecological crisis to the global changes that will occur across massive spatial and temporal scales. The integration of this new trend with the question of how dramatic works undertake the problem of scale contributes to imagining the simultaneously personal, local, national, and global effects of the Anthropocene more broadly and deeply. This global turn in ecocriticism registers as a kind of dramaturgical distorting of scale, contributing to how the theatrical works I discuss in this book thrive on the tensions and contradictions inherent in scale by destabilising the anthropocentric forms of drama or what Chaudhuri refers to as traditional drama of ideas which is limited to humanly perceptible scales. The plays and performances imagine across divergent temporal and spatial scales in light of planetary environmental realities, through productively incorporating human-nonhuman scales.

At the outset of ecocriticism in the 1990s, Cheryll Glotfelty acknowledged that, in a context of ecological systems and relationships, literature – and here theatre and performance – "does not float above the material world in some aesthetic ether, but, rather, plays a part in an immensely complex global system, in which energy, matter, and ideas interact" (xix). Glotfelty emphasises the global potential of ecocritical theory and expects it to move towards a "more interdisciplinary, multicultural, and international" becoming (xxv). In "Grounding Anotherness and Answerability through Allonational Ecoliterature Formations" (2006), Patrick Murphy also analysed the main characteristics of the global and "allonational" phase of ecocriticism and predicted the inevitability of this challenge through the exploration of texts and literatures from nations other than US-based groups.[6] This global imperative, appearing first in 2000 (Slovic, "Third Wave" 4–7), is accordingly called "ecoglobalism" by Lawrence Buell, "eco-cosmopolitanism" by Heise, and the "third wave" of ecocriticism by Joni Adamson and Scott Slovic.[7] Adamson and Slovic sketch an ecocritical trend that "recognizes ethnic and national particularities and yet transcends ethnic and national boundaries" while at the same time exploring "all facets of human experience from an environmental viewpoint" (6–7). In other words, this globalist turn in ecocriticism covers a broad spectrum of interactions between humans and the physical world, offering a more appropriate and responsive way to address our planetary predicament.

The unpredictability and complexity of the Anthropocene challenges the first principles of ecocriticism and demands a global approach to phenomena such as climate change, which necessarily incorporate different scales, perspectives, and narratives. In his article "Some Climate Change Ironies: Deconstruction, Environmental Politics and the Closure of Ecocriticism" (2010), Clark argues that only oblique references to climate change are made in various anthologies of ecocriticism: "while climate change is prominent in contemporary environmental writing and science fiction, eco-criticism itself rarely directly addresses the topic in its interpretations of literature and culture" (139). With regard to the importance of the question of scale in the current age of the Anthropocene, the concepts of "eco-cosmopolitanism," "deterritorialization" (Heise, *Sense of Place*), "loops" (Latour, "Facing Gaia"), "derangements of scale" (Clark, "Scale"), and "hyperobjects" (Morton, *Hyperobjects*) are apposite: they indicate the necessity of ecocritical thinking that is not only responsive to environmental problems at a local level but also to the implications of the spatial and temporal diffusion of anthropogenic environmental changes. In her article "Globality, Difference, and the International Turn in Ecocriticism" (2013), Heise addresses the importance of "new theories of planetarity at the intersection of culture and environment" as suitable tools for sketching a map of the realities of the Anthropocene (641–2). Addressing Glotfelty's definition of ecocriticism as having "one foot in literature and the other on land" (xix), Heise asserts that the common criteria for the importance of place-attachment in ecocriticism are defined as "special closeness, cognitive understanding, emotional attachment, and an ethic of responsibility and 'care'" (*Sense of Place* 33). However, questions about the naturalness or arbitrariness of national and cultural ties to the environment has been raised by many ecocritics and, in fact, are the point at which "place-oriented" environmentalist writers and postmodern ecocritics diverge. Rather than rejecting the differences and diverse ways of life associated with human embeddedness in particular localities, Heise calls for "the urgency of developing an ideal of 'eco-cosmopolitanism,' or environmental world citizenship" (10) and emphasises the necessity of thinking cultural and ecological differences while considering their connectedness in order to keep up with the pace of accelerating global ecological changes. In this way, theory of "deterritorialization," which Heise is borrowing from Gilles Deleuze and Félix Guattari implies that "the increasing connectedness of societies around the globe entails the emergence of new forms of culture that are no longer anchored in place" and, instead, reconfigured at other scales (Heise, *Sense of Place* 10). Such a deterritorialized vision sketches a cosmopolitan pattern of community and a network of complex biological and cultural interactions shared in "global risk scenario[s]" of climate change and biodiversity and species endangerment.[8]

While the conceptual idea of eco-cosmopolitanism broadens the environmental imagination, the emphasis on a sense of the planet as a whole is not broad enough. Through the use of modern global visualisation systems such as the Google Earth database, Heise asserts that "ecologically based advocacy on behalf of the nonhuman world as well as on behalf of greater socioenvironmental justice" are based "no longer primarily on ties to local places but on ties to territories and systems that are understood to encompass the planet as a whole" (*Sense of Place* 10). In his examination of the deficiency of the meaning of the world as a whole, Jeffrey J. Cohen in *Prismatic Ecology: Ecotheory beyond Green* (2013) argues: "Space is as multiple as it is disorderly. Moving beyond the near-to-hand and pastoral . . . locals that are the focus of much environmental criticism requires emphasizing the cosmos in eco-cosmopolitanism – yet not in the classic sense of a tidy and beautiful whole (Greek kosmos means 'order, ornament')" (xxiii). To reemphasise the "cosmos" in eco-cosmopolitanism and so balance Heise's approach, I draw on Latour's compositionism. Latour's theoretical stance of compositionism offers a more critical and capacious outlook as it "takes up the task of searching for universality but without believing that this universality is already there, waiting to be unveiled and discovered" ("Manifesto" 474). He argues that the idea of compositionism is grounded upon a "common world" that is built from "utterly heterogeneous parts that will never make a whole, but at best a fragile, revisable, and diverse composite material" (474). To describe this common world, which resists any monolithic structure, Latour coined the term *kakosmos*, meaning "a horrible and disgusting mess" (481) rather than a harmonious sphere. It is, in fact, an "entangled pluriverse" or disordered, discontinuous, and heterogeneous cosmos (481), which is composed of humans, nonhumans, agentic objects, materials, and forces, while harmony between the microcosm and macrocosm is entirely lost.

In that respect, Latour argues that the image of the spherical globe must change from a pre-given and closed sphere to some sort of loop, in such a way that a growing number of loops have slowly and collectively encircled the Earth. In relation to Latour's actor-network theory, the notion of universality gives precedence to a broad and incalculable composition of network of loops and connections among localities rather than "the utopia of the [spherical] Globe" ("Facing Gaia" 93). I find Latour's theory of a network of loops in place of the globe very encouraging for its imagining of an ecological assemblage, a composition, in which stories of human and nonhuman agency "have been melted together and morphed into strange new geological – . . . *geostorical* – forces" (77). I deploy these frameworks of thought in an attempt to examine the expanded spatiotemporal scope of Lewis's *Song of Extinction* in Chapter 4. I will demonstrate how the play sketches the thought of extinction and the end of the species as a parallel event within an eco-cosmopolitan structure in order to show an ecologically interconnected intimacy across a variety of personal, local, and global forms of loss. To gain a broader perspective of the agential feature of extinction and its distribution across intimate and global scales, I show that the play is structured in a looping way as it weaves each extinction scenario thread by thread until all extinction scenarios are connected. Within this network, the nonhuman species' extinction is no less important than the loss of a community, and might be as intimate as a human being's loss of a loved one. These interconnected narratives provide alternatives to the current Anthropocene mass extinction that offers a slow but massively destructive threat to both the human and the nonhuman world. For Latour, through the geostory of each loop, we become "sensitive," "responsive," and "responsible" to our mutual constitution and "to the fragile envelopes we inhabit" by being more deeply enmeshed in the loops' complex ongoing stories (95). Hence, shifts in perception and logic caused by global environmental changes demand new sensitivities. A reflection of all these possibilities in the ecocritical reading of a theatrical work contributes to new awareness of what it is to live on a geological time scale.

Moreover, in *Imagining Extinction: The Cultural Meanings of Endangered Species* (2016), Heise goes further and expands her argument by introducing the cultural enterprise of "multispecies cosmopolitanism" and "multispecies justice" as a response to the conceptual challenges of the Anthropocene. Heise questions Dipesh Chakrabarty's "negative universalism,"[9] which addresses a universal human species agency by subjecting everyone equally and undifferentially to the risk of global ecological changes. Instead, she insists on the complex processes of assembly of various cultures and histories through which a collective "we" might be experienced in engagement with nonhuman species and forces (Heise, *Imagining Extinction* 225). For Heise, "the planetary storytelling mode of speculative fiction" plays a key role in imagining new narratives of multispecies justice and disputing human exceptionalism (204). This narrative strategy – which I discuss in more detail in my reading of Hall's speculative play *A Brief History of Human Extinction* in Chapter 5 – "quite explicitly moves from humans' extermination of other species to a vision of multispecies contracts, constitutions, and communities on other planets, thereby contributing to the imaginative and narrative possibilities . . . of a cosmopolitanism that works through such assemblies toward multispecies justice" (Heise, *Imagining Extinction* 18). For Heise, such multispecies approaches offer key opportunities for rethinking the concept of the human, the idea of justice, and the convergence of different cultural framings and perspectives.

In "Questions of Scale: The Local, the National and the Global" (2011), Clark has built on Heise's theory of eco-cosmopolitanism; he does this by challenging the concepts of "bioregionalism" and "methodological nationalism" and the ecocritical readings they inform (136).[10] The biological disruption and ecological displacement caused by humans exceed local bioregions because they affect geological and planetary scales. Clark's book, *Ecocriticism on the Edge* (2015), has led the way by proposing shifts in ecocritical thinking that incorporate the planetary scale of the Anthropocene. Clark's idea of "derangements of scale," which he defines in relation to the difficulty of mapping the temporal and spatial scales of our current planetary conditions, examines the restrictions of human-scale perception and describes the problem of representation. Clark states that while cartographic concepts of scale imply some measure of correspondence between imaginary distances on a map and real ones on the earth's surface and are founded upon a "smooth zooming in and out," climate change – as it is induced by global-scale human interference in the planet's circulatory system – involves a map of the whole earth (*Edge* 71). However, while the scale of devastating change involves the entire globe, "when it comes to relating the threat to daily questions of politics, ethics or specific interpretations of history, culture, literature or other areas, the map is often almost mockingly useless" (71). For Clark, non-cartographic concepts of scale relating to climate change are bewildering and mystifying to calibrate, involving incalculable scale effects (72).

This theory provokes a rethinking of the interactions among bodies and spaces in the complicated and incalculable situation that Clark nicknames "Anthropocene disorder" because it specifies "a feeling of a break-down in the senses of proportion and of propriety when making judgements" (*Edge* 145). The scale effects of climate change disturb immediate perception and familiar patterns of knowledge at the personal and local scales. Clark's interpretation of the "carbon footprint" metaphor contributes to imagining the vastness of such violence. Clark argues that the carbon footprint is a metaphor or a "peculiar catachresis" that translates the quantity of carbon emissions for which an individual is directly responsible into a sensory image – a "footprint" that thereby translates the scientific data "into something intelligible in terms of the individual human body and its immediate surroundings" ("Phenomenology" 285–6). However, in reality, our footprint cannot be limited to a particular sensory perception; rather, it is an impact composed of other forces in the complex network of ecology. For Clark, this "largely benumbed recognition" of the reality of climate change and its scale effects is outside the scope of phenomenology or any system of thought bound to "the scale of individual life" and formations of localism (286–7). Such incongruent yet intertwined multi-scalar perspectives – that is, derangements of scale – disrupt any distinction between cause and consequence, subject and object, human history and natural history, and intention and result. Although it might seem that the slow pace and temporal and spatial dimensions of global ecological changes exceed the medium of theatrical representation, Clark's concept of derangements of scale provides an innovative model for the study of scales through theatre and performance. While traditional dramatic forms consider merely the personal scale of human characters, along with the local, national, and historical contexts in which the characters are embedded, a deranged scalar perspective "zooms out" of the story and displaces human characters from the centre to the margin. For example, the scale of agencies in Bilodeau's *Sila* involves a collection of divergent perspectives – including those of humans, nonhuman animals, natural forces, the Arctic, climate, and matter – on the challenge of climate change, disposing the shape and force of ecological thinking towards derangements of scale. Applying this different definition of scale to the spatial and temporal space of theatrical performance means reaching beyond anthropocentrism and zooming out of the humanist and individualist bias in theatre.

This book also considers a scale model that emerges from Morton's object-oriented ontology. Morton addresses questions similar to those of new materialists, asking us "to rethink matter at the most profound scale we have yet imagined" ("Materialism Expanded" 2). In *Hyperobjects: Philosophy and Ecology after the End of the World* (2013), he introduces us to real and autonomous but uncertain, uncontrollable, and hyper-dimensional nonhuman objects, including the biosphere, the solar system, global warming, nuclear materials on Earth such as plutonium or uranium, and "the very long-lasting product of direct human manufacture, such as Styrofoam or plastic bags, or the sum of all the whirring machinery of capitalism" (1), along with the Covid-19 pandemic.[11] These nonhuman "hyperobjects" are beyond the limited scope of our perception, and so we perceive them temporally and through their local manifestations. They are "massively distributed in time and space relative to humans" (1), are the primary cause of "the end of the world" (2), have their own agency, and are able "to make decisive contact with humans" (5). Some specific features of eco-ontological hyperobjects that are useful for my project are the hyperobjects' ability to engage the questions of scale and agency and the related concept of deep time. This book's third chapter in particular considers the deployment of hyperobjects as a tool for ecocritical thinking and analysing the materiality of the spatial-temporal scale in the two Australian plays. By questioning emissions-intensive activities and lifestyle and gesturing toward reciprocal relations between human and nonhuman materiality, the plays I consider in this chapter foreground the intrusion of large-scale matter into the local and personal scale of individual life and the merging of human timescales and geological timescales. The plays' narratives of the interactions between individual action and global effect, which is imagined as a distributed and large-scale form of materialism or hyperobject, pose profound challenges to our ways of thinking about human subjectivity, time, space, and the end of the World (Nature).

Drawing upon this new paradigm in ecocriticism, this book's ecocritical readings of selected theatrical works derive from a deranged and multi-perspectival conception of scale and explore how meaning is produced in a complicated exchange of dynamic matters and forces. In doing so, the focus on the eco-cosmopolitan vision of ecocritical theory opens a new way of thinking about the entanglements of multitudes of human and nonhuman bodies across geographical, racial, and social boundaries, informing the material trend of ecocriticism in my reading of the plays and performances. In the context of the Anthropocene, where the effects of global ecological changes are distributed over temporal and physical scales and across flows of matter and energy, we see ourselves as displaced by and thoroughly entangled with a dynamic network of agencies. Hence, I combine insights from these two trends of ecocriticism in order to imagine agency across scales and shape a framework for analysing these entanglements in theatre and performance. The plays and performances I discuss in this book also contribute to representing such complex ecological thinking by narrating stories of multispecies scales, temporalities, and agencies and, accordingly, unfolding a non-anthropocentric trajectory for theatre and performance in the Anthropocene.

Towards a material ecocritical turn in theatre studies: the challenge of agency and narrativity

Ecocriticism's evolution in its fourth wave should be considered a complement to, rather than a replacement of, the other – the third and earlier – waves of ecocriticism. In *ISLE* in 2012, Slovic heralded a "fourth-wave material trend in ecocriticism" ("Editor's Note" 621). According to Slovic, material ecocriticism brings to the fore "the fundamental materiality (the physicality, the consequentiality) of environmental things, places, processes, forces, and experiences" (619). In the same year that *ISLE* published a special cluster titled *Material Ecocriticism: Dirt, Waste, Bodies, Food, and Other Matter*, edited by Dana Phillips and Heather I. Sullivan, Serenella Iovino and her Turkish colleague Serpil Oppermann introduced material ecocriticism in their article "Material Ecocriticism: Materiality, Agency, and Models of Narrativity" (2014). This new paradigm of ecocriticism is founded upon work in new materialism by Latour, Barad, Bennett, Alaimo, and David Abram, along with work produced in other theoretical fields, "ranging from biosemiotics to the ecology of mind, from ecological postmodernism to posthumanism, [and] from 'thing theory' to object-oriented ontology (OOO)" (Iovino and Oppermann, "Stories Come to Matter" 10). Iovino and Oppermann argue that the emergence of a "neo-materialist renaissance" in the humanities and social sciences "comes after a period of dismissal of materiality as the main result of the so-called 'linguistic turn,' namely, the view that language constructs reality" ("Material Ecocriticism" 75–6). In response to the Anthropocene, the material wave of thought in ecocriticism has contextualised agency in a new way: emphasis is placed, here, on the "agentic capacities" of the nonhuman, material world (Phillips and Sullivan 447) in opposition to traditional attributions of agency only to humans according to philosophical concepts of "free will" and intentionality (Dürbeck, Schaumann, and Sullivan 121). In light of this new approach, humans are no longer understood to be at the centre of the world or in one-way subject-object relationships, but instead seen as interacting in an open-ended and vital process of reciprocity. They exchange energy, agency, and information horizontally rather than hierarchically, with manifold nonhuman bodies, landscapes, matters, bacteria, chemicals, and other unknown "actants."

In the introduction to their edited book *Material Ecocriticism* (2014), Iovino and Oppermann outline the field's conceptual argument as follows:

> the world's material phenomena are knots in a vast network of agencies, which can be 'read' and interpreted as forming narratives, stories. Developing in bodily forms and in discursive formulations, and arising in coevolutionary landscapes of natures and signs, the stories of matter are everywhere: in the air we breathe, the food we eat, in the things and beings of this world, within and beyond the human realm. All matter, in other words, is a 'storied matter.'
>
> ("Stories Come to Matter" 1–2)

Given this new perspective on matter's role in cultural analysis, material ecocriticism contributes to the examination of "matter both *in* texts and *as* a text" (2). Two points crucial to this definition of material ecocriticism are that matter is generative and has agency, and that matter's activities and configurations are "storied" and so "can be the object of a critical analysis aimed at discovering [matter's] stories, its material and discursive interplays" (79). In this way, the material trend in ecocriticism inquires not only about how the agency of nonhuman forces is represented and reflected in texts, but also about the way in which an agentic force itself narrates the story of its materiality through interaction with other forces.

This view of the world as a material phenomenon and the emphasis on the agency of matter, and its capacity to present its stories in a field of ecological interconnection generate productive models for reconceptualising human-nonhuman relationships in theatre studies. Specifically, the innovative ideas of "material narrativity," "narrative performance," and "storied matter" inform my ecocritical reading of the selected theatrical works in this book, as they challenge the narrowly anthropocentric bias of theatre by depicting a materially interconnected world in which the stories of human and nonhuman players emerge through one another. This understanding of narrative performance helps show how the plays and performances materialise new stories emerging from dynamic interactions among human imagination, language, and creativity and the narrative agency of the nonhuman world, including bodies, nonhuman species, things, and toxic substances (Chapter 2); weather, climate, and energy (Chapter 3); climatic forces, landscapes, melting glaciers, and polar bears (Chapter 4); fungi and nonhuman creatures (Chapter 5). The plays and performances give voice to the complexity of such corporations by identifying all matter, from cells to planetary transformations acting as a narrative of matter beyond human intention on different scales: the quantum level, the local, the global, and the biosphere.

Material ecocritical theory overlaps with ideas of "eco-materialization" and "materialist-ecological discourse" that Chaudhuri and May present in their theatre scholarship. The "turn toward the literal," which stands against "the use of nature as a metaphor" (Chaudhuri, "Ecological Theater" 29), shows that not only humans but also all other nonhumans can affect change in the world. This perspective also gives us the opportunity to discover the insights a work of art can thematically and performatively "provide into our material embeddedness and enmeshment in and with the more-than-human environment" (Arons and May, *Performance and Ecology* 2–3). What drives the impetus for this turn toward the literal and material expression of theatrical language, as May argues, is theatre's "inherent communality" ("Greening the Theater" 85) and "inherent reciprocity" (*Earth Matters on Stage* 3). I would also add that theatre's narrative devices, live performance, and capacities to replay different spatial-temporal frames and to embody alternative stories can communicate science in an art form. In its immediacy and flexibility, theatre has the potential to be a physical place that represents the narrativity, productivity, and dynamism of the material world. Further, theatre's collective and immersive space – which incorporates diverse humans, nonhumans, places, cultures, and objects as active participants – seems to embody aesthetically what Heise calls "the productive conceptual tension between humans' agency as a species and the inequalities that shape and constrain the agencies of different kinds of humans, on one hand, and between human and nonhuman forms of agency, on the other" ("Planet, Species, Justice" 6). Thinking about such tensions provides the opportunity to reflect on how the dramatic form and practices of performance redefine both the human and the agency of the nonhuman physical world. A concrete example of the embodiment of that tension in theatre can be seen in Bilodeau's *Sila*, which I examine in Chapter 4. The play addresses the challenge of dramatising climate change by foregrounding an ensemble of human and nonhuman stories of loss. While the play considers the geological agency of humans in terms of its planetary impact, this theatrical work highlights that human agency is determined and defined by socioeconomic and cultural circumstances, indigenous status, generation, and material exchanges and interactions. The play connects various humans to the polar bears, the Arctic landscape, the climate, and material flows to reimagine and revitalise ecological changes within a multispecies, agential, and planetary conception of scale.

In order to conceptualise the material nature of the theatrical environment, I turn to new modes of critical inquiry which might aid in considering how a play is produced by and produces a performative space of entangled forces, and in thinking about the implications of this collaboration for the idea of human agency. By drawing on Haraway's ideas of "sympoiesis," "ongoingness" and "becoming-with," I show how this theoretical framework provides new concepts for ecocritical reflection in theatre. Like Latour's idea of composition, the concept of sympoiesis is "a word for worlding-with, in company" while humans and nonhumans "interpenetrate one another, loop around and through one another, eat each other. . ., and thereby establish sympoietic arrangements" (Haraway, *Staying with the Trouble* 58). This picture of vitality and vibrancy contributes to understanding how meaning emerges through a world of collaboration and composition. In this respect, Haraway passionately advocates changing how we think about and narrate the current "troubling," "disturbing," and "mixed-up" times of mass extinction and climate change through proposing a time-space called the Chthulucene as an alternative name for our present epoch (1). Haraway describes the root and meaning of this novel term in this way: "It is a compound of two Greek roots (*khthôn* and *kainos*) that together name a kind of timeplace for learning to stay with the trouble of living and dying in response-ability on a damaged earth" (2). The concept is rooted in the webbed and "tentacular" stories of multispecies collaboration in n-dimensional time-spaces (33) and opens up ways of engaging with the loopy process of sympoiesis.[12] I deploy this materialist mode of ecological thinking to propose a non-anthropocentric dramaturgy of sympoietic thinking and worlding in theatre studies. The plays and performances in Chapter 5 in particular resonate firmly with Haraway's Chthonic through demonstrating a theatrical space of multispecies collaboration which challenges bounded individualism and human exceptionalism and, to use Haraway's expression, generates an ethics of becoming-with and making kin in heterogeneous temporalities and spatialities. The works consist of several narratives related to humans and nonhumans. In Hall's *A Brief History of Human Extinction*, fungi change into agential forces and are entangled with the last living sea otter, generating tentacular webs of destructive relations in an unpredictable way beyond human imagination. The play's sympoietic narratives imply a dynamic performance of material communications and meanings as well as unexpected configurations and entanglements according to the changing environment. Matreyek's *Infinitely Yours* also designs conditions for multispecies flourishing and cultivation of collective care in the face of the earth's mass extinction crisis. Through a practice of living and dying together on a damaged planet, the plays and performances invite us into thinking of ongoingness and staying with the troubled and muddled present.

For Haraway, a key feature of staying with the trouble and ongoingness in the Chthulucene is string figures (SF), including science fiction and speculative fabulation. As Haraway argues, SF is one definition of worlding. It is "storytelling and fact telling; it is the patterning of possible worlds and possible times, material-semiotic worlds, gone, here, and yet to come" (31). This theoretical method harmonises well with Heise's recent argument that the terraforming nature of speculative fiction and its "capacity to cast the present as a future that has already arrived" (*Imagining Extinction* 203) make it a useful mode for examining the fact that the Anthropocene itself can be understood as "a science fiction trope." The Anthropocene is grounded on "the idea of a planet terraformed by humans" (217) – wherein the division between deep time and human time has been disrupted – while this idea is connected to similarly "futuristic projects of 'de-extinction'" (18). The collaborative space of Hall's *A Brief History of Human Extinction* in Chapter 5 integrates performance art with science fiction. This fusion might be considered a different ecocritical narration and performance of the enormity of the human-made extinction crisis that gives new insight into the Anthropocene as a science fiction trope and a framework for reimagining the future. Rather than envisioning the future planet as a dead end, this speculative performance dramatises the ideas of worlding and fact-telling inherent in science fiction to think more broadly and productively about the future possibilities of the changing planet that humans might inhabit differently in the Anthropocene.

A similar approach to theatre as a materialist process of performativity might be imagined through Barad's challenge to representationalism and individualism, revealing a theatrical work as an ecological practice in which heterogeneous human and nonhuman actors are inextricably entangled with each other and generate a wide range of meanings and effects. In *Meeting the Universe Halfway* (2007), Barad develops a "diffractive" methodology that provides new insights into the concept of performativity and how the world comes to matter. Borrowing from Haraway, "diffraction" can serve as an alternative and, in fact, a counterpoint to "reflection." Barad elaborates on the term to determine how diffraction acts "as a tool of analysis for attending to and responding to the effects of difference" (72). The concept provides an innovative methodological approach for rethinking the Anthropocene, in which different agential forces such as humans, nonhuman animals, objects, and weather have their own narrativity; also, by being relationally, processually, and differentially engaged in a material entanglement, these forces can produce different effects. Resonating with Lavery's description of theatre as a form of ecocritical doing, Barad's theory of diffractive and "performative alternatives to representationalism" shifts the focus from "questions of correspondence between descriptions and reality to matters of practices, doings, and actions" by highlighting issues of materiality and agency ("Posthumanist Performativity" 802). The theatrical works I study here enact performative alternatives to the conventional epistemological forms of representationalism and humanism through productively troubling human supremacy and anthropocentric knowledge practices and fostering a collaborative space in which human histories, agencies, values, desires, and expectations are integrated with broad networks of forces.

To explicate this dramaturgical praxis of material expression and performative approach to agency and representation in theatre, the key idea of "intra-activity" constitutes a useful conceptual framework. Against the common term "interaction," which suggests that the world consists of autonomous, self-referential, and pre-existing entities or relata, the concept of intra-action – which can be regarded as an extension of Haraway's sympoiesis and tentacular thinking – necessitates a dynamism of forces and proposes a communicative and agential mode of exchanging and becoming (Barad, *Meeting the Universe* 150). Understood as an intra-active practice, a play challenges common narratives comprised of pre-existing entities and foregrounds the agency of humans, nonhuman animals, weather, landscape, carbon emissions, and many other chemical, physical, biological, and geological formations, along with the complexities of their intersections, through open-ended and generative processes and embodiments of mutual and ongoing becoming. This materialist account of performativity is also concerned with Barad's idea of iterative reconfigurations of "spacetimematter." This concept underpins the following assertion that an understanding of agential and spatial intra-activity helps to reconfigure personal and local boundaries as well as boundaries between humans and nonhumans. For Barad, intra-active processes are the very forces through which time and space are produced. Reflecting on the distortion of time and space in the context of a changing climate, Barad argues that time is neither "an exterior parameter" nor "a succession of evenly spaced intervals available as a referent for all bodies," and that "space is not a collection of preexisting points set out as a container for matter to inhabit" (234). The different scales of the local, national, and global are, according to Barad, "topological matters, intra-actively produced through one another" (246). This idea indicates that understanding the dynamics of unprecedented scales of global ecological risk such as climate change requires a different kind of analysis: one that responds to the intensity of the phenomenon and is not restricted to geometrical measurements of scale, which are founded on a hierarchical relationship among local, national, and global. In my ecocritical reading of Bilodeau's *Sila* in Chapter 4, this dynamic intra-activity evokes a non-anthropocentric scale of knowledge about the world. The play features large-scale anthropogenic climate change from the perspectives of human and nonhuman actors whose stories end up converging. As melting ice glaciers replace the paradisiacal and transcendental sense of "land" with "climate" catastrophe, the play necessitates acknowledging the performative intra-action between human and nonhuman agents in narrating the ecological changes. This understanding contributes to a redefinition and recomposition of the human, matter, space, and time as they come into play through intra-active relation to landscape transformation and iterative reconfigurations of boundary scales.

The nature of such processes of intra-action can also be demonstrated through thinking of a theatrical work as an "apparatus." This perspective yields a new way of seeing how human and nonhuman actors, things, matters, the elements of space and sound, digital technology, each iteration of a production, the cultural and physical contexts of each performance, audience participation, and each reading and analysis of the play are practiced and experienced as intertwined critically, dynamically, and intra-actively. Minty Donald has led the way in understanding theatre and performance through Barad's concept of the apparatus, considering a play "in terms of the essential adaptability of the apparatus – its characteristics of continual, responsive reconfiguration and mutation – offers a model for how we (humans) might 'prepare' for the future, in light of ever-changing ecological conditions" (266). In this light, a theatrical work intra-actively reconfigures itself in response to the mutable and dynamic conditions of the material world. For example, in the 2015 production of Churchill's play *The Skriker* at the Manchester Royal Exhibition Theatre, critics acknowledged that its environmental themes were even more timely and disturbing than they had been two decades earlier when the play premièred at the National Theatre in London. Influenced by Naomi Klein's *This Changes Everything: Capitalism vs. the Climate*, the 2015 production openly thematises climate change and foregrounds the Anthropocenic reception of the play on stage, providing an opportunity to explore how the cultural and theatrical positioning of a play can change intra-actively in response to ever-changing ecological conditions. This possibility of reconfiguration in theatre can also be seen in the 2016 production of Bovell's play *When the Rain Stops Falling* at the Wilma Theater, Philadelphia. Influenced by Roy Scranton's book *Learning to Die in the Anthropocene* and Nixon's idea of slow violence, the director Blanka Zizka reconfigures the production in order to spotlight and activate the play's embedded theme of the impact of climate change on humanity and to show how the play's family saga deals with the Anthropocene. These plays are lively in the possibilities they generate in each production and moment of intra-action.

While Barad asks us to think about matter through intra-active performativity and Haraway calls for composite and sympoietic life narratives in n-dimensional time-spaces, Bennett and Alaimo give an account of agential bodies and things on more familiar scales, underpinning my material ecocritical inquiries in theatre. In *Vibrant Matter* (2010), Bennett proposes an ecological ethical framework by formulating a political theory that shifts the focus of agency from human subjectivity to the "force of things" (1). In an attempt to show how each material agency can make a story and meaning, Bennett underlines the idea of "vital materialism." According to Bennett, the political goal of a vital materialism "is not the perfect equality of actants, but a polity with more channels of communication between members" (104). Invoking the work of Latour, Bennett uses the concept of "actant" in her examination of the material agency of electric grids, waste, and inanimate objects that are characterised as performative agents emerging from intra-active processes. In "On Actor-Network Theory," Latour defines the term in this way: "An 'actor' in AT [actor-network theory] is a semiotic definition – an actant –, that is, something that acts or to which activity is granted by others. It implies no special motivation of human individual actors, nor of humans in general" (373). Unlike a dichotomous view of the world, this more encompassing perspective acknowledges that we are all participants in the planetary network of material exchanges. In this definition, Latour labels the collectives of human and nonhuman actants that come together in networks of association an "actor-network" in which all the components are "in the process of exchanging competencies, offering one another new possibilities, new goals, new functions" (*Pandora's Hope* 182). Bennett then uses the term actant to propose an alternative for considering matter as transformative because its meaning, efficacy, or agency emerges through a heterogeneous "swarm of vitalities at play" (31–2). This "distributed agency" and exchange among participants is not, of course, one of uncomplicated and harmonised connection, but a complex and dark one (xi). In Chapter 2, I show that the final sequence of Churchill's *Far Away*, which sets the scene for global warfare waged by an assemblage of humans and nonhuman species and by matter and elemental forces, prompts us to think about the question of agency. The agentic materiality of the play's dark and composite environment strikingly resonates with Bennett's idea of vital materialism, providing an ontological approach to the political ecology of human and nonhuman actants. Rather than focusing on a linear pattern of causality, I focus on the play's decentring of the human and formation of parallel communicative alliances among its swarm of actants across species, nations, and the forces of nature to seek a materialist thinking of vitality and effectivity for all forms of life.

Echoing what Barad names "posthumanist performatives" and Bennet names "the agency of assemblages," I show that the selected theatrical works dramatise a non-anthropocentric practice of performativity where agency is contextualised across a composition or working group of human and nonhuman actors. It is worthwhile to add that a theatrical space itself is an ecological assemblage in the sense that it is an interconnected series of parts. So the collaboration of theatrical materials such as the actors, scenery, costumes, light, sounds of music and of words, and electrical and technical equipment with the swarm of variegated actants to which the characters refer throughout a play can more deeply shape our awareness of ecological relations. This perspective distinguishes ecocritical theatre from traditional anthropocentric theatre by problematising the relationships between human and nonhuman agency. While the conventional practice of theatre and performance is grounded upon traditional notions of agency that insist on a plan or an intention as the generative source of a subject's actions, ecocritical theatre considers the transformative function of performance when matters embody and perform agency and an exchange between two or more things generates effects and meanings.

Many of these themes are echoed in Alaimo's *Bodily Natures* (2010), which contextualises agency on a "trans-corporeal" scale. Alaimo's innovative concept of "trans-corporeality" plays a key role "in drawing the intellectual map of material ecocriticism" (Iovino, "Matter, Text, and Posthuman Ethics" 55). Trans-corporeality focuses on "the material interrelatedness of all beings, including the human" (Alaimo, *Bodily Natures* 151) and allows us to trace faint courses of interface and affect among multiple bodies across time and space. This counter-intuitive theory and Bennett's idea of vital materialism should be considered as theoretical lenses to put in conversation with the material ecocritical reading of theatre by introducing productive ways of knowing and thinking about matter and the agentic interplay between humans and nonhumans. In this way, theatre can also help us develop the material turn in ecocriticism by disassembling the anthropocentric tendencies, unearthing the nonhuman world, making palpable the entangled ecologies, and gesturing towards the vibrant and trans-corporeal nature of the material world in, and out of, the Anthropocene context.

Alaimo's conceptual model of trans-corporeal entanglements provides a foundation for my reading of Churchill's *The Skriker* as a potent example of trans-corporeality. Through the shape-shifting character of the Skriker, who materialises temporal and spatial dislocation in a fragmented eco-global space, my reading of the play considers the material interchanges between bodies (both human and nonhuman) and the wider environment. I will focus on the mutative feature of time/space, nonhuman figures, and even language as embodiments of trans-corporeality, insisting that "the human is always the very stuff of the messy, contingent, emergent mix of the material world" (Alaimo, *Bodily Natures* 11). In a trans-corporeal space of ecological risk and toxicity, manifold bodies and movements – including "material, economic, and cultural systems that are so harmful to the living world and yet so difficult to contest or transform" (18) – are negatively and dangerously affected by each other. As Heise recognises the impact of Ulrich Beck's risk society, Alaimo argues that the invisibility of trans-corporeal spaces of risk is due to "the often unpredictable and unwanted actions of human bodies, nonhuman creatures, ecological systems, chemical agents, and other actors" flowing in a "swirling landscape of uncertainty" (20). This acknowledgement has become crucial to imagining material interchanges among human and nonhuman bodies and a broader global network in which world ecology, world economy, and world policy are inseparable while also requiring scientific mediation (Alaimo, *Bodily Natures* 16).[13] One instance of such trans-corporeality in Churchill's play is through the toxic body, which is depicted through the fragmented bodies of the human characters and their constant interchange with ecological, economic, political, and material forces. Because "the substance of one's self" undergoes mutations as a result of encountering environmental hazards and risks, Alaimo argues that we need a new genre of autobiography, which she nicknames "material memoirs" (23–4). The material world "is never merely an external place but always the very substance of our selves and others" (158), making human characters in a play materialise like objects or things in connection with a global ecological network. This agential shift and new conceptualisation of the human is evident in *The Skriker*, in which the toxic environment becomes the very material of human bodies, mutating and diminishing them to some poisoned food and "dustbin." This exchange across bodily natures profoundly shifts the meaning of subjectivity and generates a material self.

Drawing connections among studies of ecological theatre, material ecocriticism, and global ecocriticism, this book argues that in an age of ecological terror and risk, an ecocritical reading of dramatic works must be engaged materially and globally in the representation of environmental uncertainty. My use of theoretical frameworks from writers like Latour, Haraway, Barad, Alaimo, Bennett, Heise, Clark, and Morton shows theatre's relevance to the cultural shifts induced by the Anthropocene and reveals complementary ways of thinking about questions of agency and scale that are important to theatre and performances studies. Integrating such new perspectives into the field of theatre contributes to disturbing our anthropocentric relationship with the environment, foregrounding the agency of the nonhuman world, shaping imaginative models of understanding agency across scales, and affecting our ecological response to shared planetary concerns such as climate change. Ecocritical theatre provides a compelling opportunity for spectators to observe themselves in the performative nonhuman world and see the material world in themselves – not just intellectually but at a personal and emotional level.

Chapter summaries

Chapter 2 brings a material ecocritical perspective to the plays *The Skriker* (1994) and *Far Away* (2000) by British playwright Caryl Churchill. When we read Churchill's plays through a material ecocritical lens, we become attuned to the plays' articulation of an agential account of earthly life and a non-anthropocentric perspective on the world, which enables us to reconfigure and rewrite our life stories in connection with a multitude of nonhuman animals, toxic substances, things, commodities, and climate forces as they occupy and alter our bodies. I examine *The Skriker*'s contribution to rethinking how nonhuman figures manifest as agents and the human body as a material entity. By moving between the human world and the underworld, the play presents a sequence of encounters between an ancient and damaged fairy called the Skriker – alongside a collection of wordless but mutative fairy figures – and two female characters. I shape my ecocritical reading around the composite figure of the Skriker whose shapeshifting and constant mutations, its frantic and fractured form of language play, and its circulation across time and space and ability to disrupt spatio-temporal logics display the materiality and performativity of the nonhuman world beyond human intentions, understandings, and agencies. My evaluation of the play's staging of agentic materiality resonates with Alaimo's new materialism and her concepts of "trans-corporeality" and "toxic bodies." Reading the play through such a lens emphasises what it would mean to consider the Anthropocene through chemical and biological transformations in addition to geological turns. *The Skriker*'s theme of environmental toxicity is embodied through fairy figures and their trans-corporeality, or material interconnectedness with human bodies as produced by collective agencies and dynamics. In making use of the language of "vibrant materialism" without depicting it onstage, *Far Away* provides a broader image of the dynamic landscape in which all entities and matter are figured with a kind of material vitality in response to the complexities of the global ecological mess. The third act of the play imagines a transformative material world under the conditions of a truly global warfare that encompasses not just all of the world's nations and animal species but all forces of nature. In this way, the play anticipates Bennett's "distributive agency," which gives voice to the "immanent vitality" distributed across human and nonhuman "actants." These plays engage with the contemporary ecological imperatives of the Anthropocene by staging vital, trans-corporeal intersections in which the human is entangled with the flows of substances and the agency of environments.

In Chapter 3, I examine Stephen Sewell's *It Just Stopped* (2007) and Andrew Bovell's *When the Rain Stops Falling* (2009) in order to explore how recent Australian theatre responds to the question of global ecological change. The plays narrate the agency and immediacy of "the end of the world" – a term used iteratively in both plays – unsettling utopian descriptions of sustainability and urging a reconsideration of our relation to matter on the largest scale. *It Just Stopped* unfolds a panicked introduction of the end of the world through the sudden interruption of fossil-fuel based energy supply and all power resources. The play represents two kinds of response to this unexpected change: denialism and apocalyptic thinking. However, as the play moves on, we can see how the footprints of climate change, as a device of high-carbon lifestyle, can be indirectly but materially detected through the unavailability of energy and all aspects of modern society. Denial of the end of the world – and indeed denial of the end of certainty in the concepts of Nature and Sustainability which are based on an aesthetic distinction between humans and nonhumans – is in part the refusal of the reality of climate change that is already pervading our daily lives. I continue to work with the idea of the end of the world in Bovell's weather-driven play. With its web of intricately overlapping connections, the play follows interconnected events in the lives of four generations of one family from 1959 to 2039, and from London to Australia. The significance of the play for understanding climate change lies in the play's growing sense that it is no longer possible to have a stable and predictable conception of weather and that human actions and family relationships are framed in light of this change. Extreme weather or the weird rain, as a real "hyperobject," embeds itself in the characters' social and psychic space, perception and ethics, extended beyond their personal and local scales of time and place, and engages even unborn generations. My ecocritical reading brings climate change into interaction with human beings to disclose their ecological contribution. My reading reconsiders the plays as forms of ecological doing that are engaged with a sense of urgency aimed at displacing the human as a central figure in discourses on the Anthropocene. In doing so, I focus on theoretical work based in object-oriented ontology, which has a tense relationship with the new materialist approach that I discuss in Chapter 2. Both approaches yield new insights into the interconnection between human and nonhuman materiality, although they show this interconnection on different scales. The most useful discussion of the scale of objects for my purpose is in Morton's theories, which deal with "expanded materialism." These theories reconfigure the issue of scale, dealing with matters that are "hyper" because they are nonlocal and atemporal. When we read the plays through this lens we see the focus of the plays shift from a human-centred drama to one that is concerned with how large-scale and agential objects – hyperobjects – reshape our definition of scale, subjectivity, lifeways, cultural and social discourses, and the end of the world.

Next, Chapter 4 considers plays that explicitly engage with the Anthropocene. I examine the materiality of mass extinction and climate change in E.M. Lewis's *Song of Extinction* (2010) and Chantal Bilodeau's *Sila* (2015). These two works, as obvious examples of Chaudhuri's new theatrical practice of "fifth wall dramaturgy," have an environmental focus at their centre. In my reading of *Song of Extinction*, I explore the confluence of human and nonhuman territories and the local and the global from the vantage point of extinction, mindful that extinction is not a singular procedure. Premiered at the Moving Arts Theatre Company in Los Angeles, the play considers a collection of intricately overlapping stories of loss to present the audience with a range of temporal and spatial scales in which the Cambodian genocide, the Bolivian ecocide, and the death of a mother in Portland in the US are juxtaposed with a global picture of Anthropocenic extinction. This interconnected intimacy across a variety of personal, local, and global forms of human and nonhuman loss involves the problem of scale and gestures towards a reconfiguration of the concept of the globe. Invoking Latour's counterintuitive idea of the "loop," I discuss the play's "deterritorialized" space, to use the term from Deleuze and Guattari that Heise reworks, as being structured in a looping way rather than as a single whole. The play's cosmopolitan network is composed of dense loops – of extinction scenarios – while all are connected in a single line of Anthropocene mass extinction and display an entangled environment of humans and nonhumans. *Sila* is set in the Canadian Arctic and represents extinction as inextricably connected to climate change; it does so by directly engaging intertwined stories and different experiences of loss and mourning. These stories are built on a multi-perspective panorama and silent narrative voices – of a Canadian First Nations family, a mother and daughter polar bear, and seemingly static and heavenly nonentities such as the Arctic landscape and ice glaciers, which are in reality dynamic forces in their own right. However, the question about scale that provides the clearest perspective on the play's staging of climate change eventuates in what Clark has identified as "derangements of scale," challenging theatre's supposed limitation to personal and national scales. In particular, I will address the ways in which the play's involvement with such uncertainty necessitates an acknowledgement of the performative "intra-action" between human and nonhuman agents in narrating ecological change. To illuminate the play's staging of complex intra-relationships, I draw on Barad's concept of "posthumanist performativity," which offers new epistemological, ontological, and ethical awareness of such material entanglements. In regard to the importance of the agentic potentiality of the Arctic landscape, I posit that the Arctic speaks through its ecology and the complex stories of humans and nonhumans, speaking and acting intra-actively as performative forces in ways that are central to the action. Further, the play's dynamic performativity – achieved through the intra-active feature of the Inuktitut word Sila (meaning breath, spirit, the air, the wind, the weather, consciousness) – fosters a materialist perspective on the world, in which our breath and awareness are interwoven into that of nonhumans and their embodied stories.

Finally, Chapter 5 examines *A Brief History of Human Extinction* (2018) by Canadian playwright Jordan Hall and *Infinitely Yours* (2020), along with *This World Made Itself* (2013), by Miwa Matreyek, an animator, director, designer, and multimedia performance artist based in Los Angeles. My ecocritical readings of the works in this chapter and Chapter 4 begin with a focus on how they engage with the issue of scale and end with a material approach to their representation of ecological relationships, accompanied by reflections on agency and narrativity. *A Brief History of Human Extinction*, jointly created by Hall and the company Mind of a Snail, critically intervenes in discourses about mass extinction by staging the interaction between the last remnants of the human species and the last living sea otter sealed inside a decrepit biodome, together with a toxic network formed by fungi. The play explores the urgency of this complex situation by integrating performance art into science fiction. This integration embodies the present and future life of human and nonhuman "critters." Drawing on Haraway's idea of multispecies "sympoiesis" and making together, I consider the play's contribution to the emergence of an ecocritical theatre in which humans are in the process of "making kin" and "becoming-with" other species, elements, and the planet. The play's representation of this era of sympoietic thinking involves breaking with human exceptionalism and embracing the new era of the Chthulucene, wherein humans and nonhumans are inextricably linked in interwoven and nonhierarchical practices while "staying with the trouble" of living and dying together on a wounded planet and thus making conditions for multispecies flourishing. Miwa Matreyek's enthralling sensory and visual performative arts mark a deliberately dramatic closure to this book as they involve profoundly and palpably staging the collaboration between art and the Anthropocene. These ecodramaturgical practices of sympoietic thinking and action integrate collage animation, dance, musical elements, installation art, film, theatrical performance, and eco-activism. The intricate narratives of the material world and humanity in Matreyek's live shadow silhouette and projected animations journey from the birth of the universe forward through the expansive history of the earth in *This World Made Itself*, eventually moving to the current state of global biosphere in *Infinitely Yours*. The rapidly evolving visual narratives – and, in particular, Matreyek's shape-shifting shadows that traverse a range of scales – enact a swiftly changing and evolutionary world and display the complex idea of deep time and its interference with human timescales. *Infinitely Yours* demonstrates this ecological displacement by highlighting how the profound impact of human activity on geological processes has generated the changes (hyperobjects) that are beginning to act on massive scales of geological deep time. Announcing our place in a complex network of co-emerging interactions, the work is an embodied illustration of what it means to live in this ever-changing world of vital materiality. This multimedia live performance work brings together all the pictures of environmental damage represented in the previous plays and so makes a fitting conclusion to this book.

In the closing, I review my findings about the selected plays and performances and also describe the connections that emerge among them. *Towards an Ecocritical Theatre* provides a framework that shows how ecocritical thought about the Anthropocene enables a non-anthropocentric perspective on theatre. Through this lens, the book advances an ecological model of theatre. This model calls for recalibrating our outlooks to acknowledge and understand the human body over multiple scales and in interaction with a range of nonhuman agents and meanings that are transformed by their encounter with the Anthropocene. Seen in this light, the theatrical works might well become forms of thought and afford a new kind of knowledge, participating in imaginative conversations on the most urgent issues of our time. This book opens up space for further scholarship on the Anthropocene-oriented ecocritical turn in theatre and performance.

Notes

1 See Sarah Lazare's "'Colonizing the Atmosphere': How Rich, Western Nations Drive the Climate Crisis" (2020).

2 The other examples include: Nick Payne's *If There Is I Haven't Found It Yet* (2009), Zoë Svendsen's *3rd Ring Out: Rehearsing the Future* (2010), *Greenland* by Moira Buffini, Matt Charman, Penelope Skinner, and Jack Thorne (2011), Duncan Macmillan's *Lungs* (2011), *Ten Billion* by Stephen Emmott and Katie Mitchell (2012), Duncan Macmillan's *2071* (2014), Tanya Ronder's *F*ck the Polar Bears* (2015), Stef Smith's *Human Animals* (2016), Churchill's *Escaped Alone* (2016), Ella Hickson's *Oil* (2016), Lucy Kirkwood's *The Children* (2016), and *Arctic Oil* by Clare Duffy (2018).

3 A major challenge facing the Anthropocene concept is that, even as it works to end the human-nonhuman binary, it frames the category of the "human" in an undifferentiated, globally abstracted sense without sufficient regard for history, economics, gender, race, and ethnicity. In the collection *Anthropocene Feminism* (2017) edited by Richard Grusin, contributors Rosi Braidotti (posthuman feminism), Claire Colebrook (counterfactual thinking and (in)difference), and Stacy Alaimo (trans-corporeality) emphasise how an Anthropocene feminist perspective offers an effective means for extending critical reflections on the concept of the "human" as a geological force. Concerning this problem, feminism is also needed to reformulate thinking of gender and power through intersecting "the scale of the personal" with a geological scale (Colebrook, "The Anthropocene Counterfactual" 1). Further, the edited book *Feminist Ecologies: Changing Environments in the Anthropocene* (2018) by Lara Stevens, Peta Tait, and Denise Varney critically considers the contemporary and future challenges for ecofeminism in light of Anthropocenic environmental issues. The collection "offers a timely examination of how ecofeminist thought, its histories and its activism intersect with global debates about environmental catastrophe as a product of global corporate capitalism and neoliberalism" (Stevens et al. 17).

4 Since the ecological theatre edition of the journal *Theater* in 1994, scholarship in the field has been the subject of a number of articles and books. Some of these works include: the 2004 and 2005 edited collections of *Nature Performed* by Bronislaw Szerszynski, Wallace Heim, and Claire Waterton and *Performing Nature* by Gabriella Giannachi and Nigel Stewart; the 2006 *Journal of Dramatic Theory and Criticism*, volume 20, issue 2; the September 2007 issue of *Theater* edited by Wendy Arons the journal of *Canadian Theatre Review: Eco-Theatre*, volume 144, Fall 2010, edited by Gary Nelson and Sheila Rabbillard; the 2012 journal of *Performance Research: On Ecology*, volume 17, issue 4, edited by Stephen Bottoms, Aaron Franks, and Paula Kramer; the 2015 book *Rethinking the Theatre of the Absurd: Ecology, the Environment and the Greening of the Modern Stage* edited by Carl Lavery and Clare Finburgh; Kirsten Shepherd-Barr's book *Theatre and Evolution from Ibsen to Beckett* 2015; the 2016 *Journal of Green Letters: Studies in Ecocriticism*, volume 20, issue 3, on "Performance and Ecology: What Can Theatre Do?" edited by Carl Lavery; ecoscenographic projects including the 2017 article "Saved from

the Scrapheap: Revealing the Creative and Ecological Potential of Societal Leftovers in Scenography" by Tanja Beer and the 2020 article "Justice Scenographics: Preparing for civilization change in a time of 'Anywheres' and 'Somewheres'" by Rachel Hann; the 2020 *Journal of Nordic Theatre Studies*, volume 32, issue 1, on "Theatre and the Anthropocene" edited by Stephen Elliot Wilmer and Karen Vedel; the 2020 book *Earth Matters on Stage: Ecology and Environment in American Theater* by Theresa May; the 2020 article "Tragedies of the Capitalocene" by Wendy Arons; and the 2020 book *Ecodramaturgies: Theatre, Performance and Climate Change* by Lisa Woynarski.

5 Notable examples include: the 2000 conference "Between Nature: Explorations on the Edge of Ecology and Performance" in the UK; "Mapping Ecocriticism across Performance" at the 2005 American Society for Theatre Research conference in Toronto; "Earth Matters on Stage: Ecodrama Playwrights Festival & Symposium" (2004, 2009, 2012, 2015, 2018); the Australasian Association for Theatre, Drama and Performance Studies 2014 conference "Restoring Balance: Ecology, Sustainability, Performance" in Wellington (New Zealand); the 2015 colloquium on Artistic Research in Performing Arts, "The Non-human and the Inhuman in Performing Arts," at Theatre Academy Helsinki (Finland); the 2015 Festival of Climate Change Related Arts, ART+CLIMATE=CHANGE, in Melbourne (Australia); the 2015 ArtCOP21, a global festival of cultural activity on climate change; the 2016 conference "Environment, Economy & Climate Change: Stages in Transition" at the University of Birmingham (UK); the 2016 Performance Studies International conference "Performance Climates" at the University of Melbourne (Australia); the Canadian Association for Theatre Research 2017 conference "Performing the Anthropocene: Setting the Stage for the End of the World" in Toronto (Canada); and the International Federation for Theatre Research 2021 conference "Theatre Ecologies: Environments, Sustainability, and Politics."

6 The founding of ASLE-affiliated organisations in the UK, Canada, Korea, India, Germany, Australia, and New Zealand, as well as of the European Association for the Study of Literature, Culture and Environment (EASLCE), are proof of the international expansion of ecocriticism (Heise, "Hitchhiker" 505; Oppermann, "Transnationalization" 415).

7 Lawrence Buell, "Ecoglobalist Affects" (2007); Ursula Heise, *Sense of Place and Sense of Planet*; Joni Adamson and Scott Slovic, "Guest Editors' Introduction."

8 Ursula Heise argues that while the ecological scenarios of climate change and mass extinction are informed by scientific data, they are caught in webs of cultural frameworks of thought and value (*Imagining Extinction*).

9 See Dipesh Chakrabarty's "The Climate of History: Four Theses" (2009).

10 As Timothy Clark writes, "methodological nationalism" is taken from A.D. Smith and used by Ulrich Beck. In his critique of "methodological nationalism," Beck argues: "Globality means that from now on nothing which happens on our planet is only a limited local event; all inventions, victories and catastrophes affect the whole world, and we must reorient and reorganize our lives and actions, our organizations and institutions, along a 'local-global' axis" (qtd. in Clark, *Cambridge Introduction* 131).

11 In her article "Hyperobject Covid-19" (2020), Serenella Iovino addresses the pandemic "as a very particular object: a hyperobject" (n. pag.).

12 According to Donna Haraway, "[t]he tentacular are not disembodied figures; they are cnidarians, spiders, fingery beings like humans and raccoons, squid, jellyfish, neural extravaganzas, fibrous entities, flagellated beings, myofibril braids, matted and felted microbial and fungal tangles, probing creepers, swelling roots, reaching and climbing tendrilled ones. The tentacular are also nets and networks, IT critters, in and out of clouds" (*Staying with the Trouble* 32).

13 Stacy Alaimo's concept of trans-corporeality takes a stance comparable to that of Ursula Heise's theory of eco-cosmopolitanism, in that both embody a shared vision of "environmental world citizenship" and reimagine the human/nonhuman relationship in global ecological, economic, political, and social networks.

References

Adamson, Joni and Scott Slovic. "Guest Editors' Introduction: The Shoulders We Stand On: An Introduction to Ethnicity and Ecocriticism." *MELUS* 34.2 (Summer 2009): 5–24. Print.

Alaimo, Stacy. *Bodily Natures: Science, Environment, and the Material Self.* Bloomington: Indiana UP, 2010. Print.

———. "Sustainable This, Sustainable That: New Materialisms, Posthumanism, and Unknown Futures." *PMLA: Publications of the Modern Language Association of America* 127.3 (2012): 558–64. Print.

———. "Your Shell on Acid: Material Immersion, Anthropocene Dissolves." *Anthropocene Feminism*. Ed. Richard Grusin. Minneapolis: Minnesota UP, 2017.89–120. Print.

Arons, Wendy. "Introduction to Special Section on 'Performance and Ecology'." *Theatre Topics* 17.2 (2007): 93–4.

———. "Tragedies of the Capitalocene." *De Gruyter Mouton* 8.1 (2020): 16–33. <https://doi.org/10.1515/jcde-2020-0003>.

Arons, Wendy and Theresa J. May. Eds. *Readings in Performance and Ecology*. New York: Palgrave Macmillan, 2012. Print.

Barad, Karen. *Meeting the Universe Half Way: Quantum Physics and the Entanglement of Matter and Meaning*. Durham, NC: Duke UP, 2007. Print.

———. "Posthumanist Performativity: Towards an Understanding of How Matter Comes to Matter." *Journal of Women in Culture and Society* 28.3 (2003): 801–31. Print.

Beck, Ulrich. *Risk Society: Towards a New Modernity*. Translated by Mark Ritter, London: SAGE Publications, 1992. Print.

Beer, Tanja. "Saved from the Scrapheap: Revealing the Creative and Ecological Potential of Societal Leftovers in Scenography." *Performance Research* 22.8 (2017): 107–14. Print.

Bennett, Jane. *Vibrant Matter: A Political Ecology of Things*. Dunham and London: Duke UP, 2010. Print.

Bilodeau, Chantal. *Sila (The Arctic Cycle)*. Vancouver: Talonbooks, 2015. Print.

Bottoms, Stephen, Aaron Franks, and Paula Kramer. Eds. "On Ecology." *Performance Research* 17.4 (2012): 1–4. <www.performance-research.org/past-issue-detail.php?issue_id=62>.

Bovell, Andrew. *When the Rain Stops Falling*. Sydney: Currency P, 2009. Print.

Buell, Lawrence. "Ecoglobalist Affects: The Emergence of U.S. Environmental Imagination on a Planetary Scale." *Shades of the Planet: American Literature as World Literature*. Eds. Wai Chee Dimock and Lawrence Buell. Princeton: Princeton UP, 2007.239. Print.

Chakrabarty, Dipesh. "Brute Force." *Eurozine*, 7 Oct. 2010. Web. 10 Apr. 2016. <www.eurozine.com/articles/2010-10-07-chakrabarty-en.html>.

———. "The Climate of History: Four Theses." *Critical Inquiry* 35.2 (Winter 2009): 197–222. Print.

Chaudhuri, Una. "Anthropo-Scenes: Staging Climate Chaos in the Drama of Bad Ideas." *Twenty-First Century Drama: What Happens Now*. Eds. Siân Adiseshiah and Louise LePage. London: Palgrave Macmillan, 2016.303–21. Print.

———. "Research Theatre." *Research Theatre, Climate Change, and the Ecocide Project*. Eds. Una Chaudhuri and Shonni Enelow. New York: Palgrave Macmillan, 2014.1–21. Print.

———. *Staging Place: The Geography of Modern Drama*. Ann Arbor: The University of Michigan P, 1995. Print.

———. "The Fifth Wall: Climate Change Dramaturgy." Web blog post. *HowlRound*. HowlRound, 17 Apr. 2016. Web. 10 May 2016. <http://howlround.com/the-fifth-wall-climate-change-dramaturgy>.

———. "'There Must Be a Lot of Fish in that Lake': Toward an Ecological Theater." *Theater* 25.1 (1994): 23–31. Print.

Chaudhuri, Una and Shonni Enelow. *Research Theatre, Climate Change, and the Ecocide Project*. New York: Palgrave Macmillan, 2014. Print.

Churchill, Caryl. *Far Away*. London: Royal Court, 2000. Print.

———. *The Skriker*. London: Cox & Wyman Ltd, 1994. Print.

Clark, Timothy. *Ecocriticism on the Edge: The Anthropocene as a Threshold Concept*. London: Bloomsbury Academic, 2015. Print.

———. "Questions of Scale: The Local, the National and the Global." *The Cambridge Introduction to Literature and the Environment*. Cambridge: Cambridge UP, 2011.130–42. Print.

———. "Scale." *Telemorphosis: Theory in the Era of Climate Change, Volume 1*. Ed. Tom Cohen. Ann Arbor: MPublishing, Open Humanities P, 2012.148–66. Web. 14 July 2015. <http://dx.doi.org/10.3998/ohp.10539563.0001.001>.

———. "Some Climate Change Ironies: Deconstruction, Environmental Politics and the Closure of Ecocriticism." *The Oxford Literary Review* 32.1 (2010): 131–49. Print.

Cless, Downing. *Ecology and Environment in European Drama*. New York: Routledge, 2010. Print.

Cohen, Jeffrey Jerome. "Introduction: Ecology's Rainbow." *Prismatic Ecology: Ecotheory beyond Green*. Ed. Jeffrey Jerome Cohen. Minneapolis: Minnesota UP, 2013.xv–xxxv. Print.

Colebrook, Claire. "We Have Always Been Post-Anthropocene: The Anthropocene Counterfactual." *Anthropocene Feminism*. Ed. Richard Grusin. Minneapolis: Minnesota UP, 2017.1–20. Print.

Crutzen, Paul and Eugene F. Stoermer. "The 'Anthropocene'." *Global Change Newsletter* 41 (2000): 17–8. Print.

Donald, Minty. "The Performance 'Apparatus': Performance and its Documentation as Ecological Practice." *Green Letters: Studies in Ecocriticism* 20.3 (2016): 251–69. Print.

Dürbeck, Gabriele, Caroline Schaumann, and Heather I. Sullivan. "Human and Nonhuman Agencies in the Anthropocene." *Ecozon@* 6.1 (2015): 118–36. Web. 20 July 2016.

Giannachi, Gabriella and Nigel Stewart. "Introduction." *Performing Nature: Explorations in Ecology and the Arts*. Gabriella Giannachi and Nigel Stewart. Eds. Oxford: Peter Lang, 2005.19–62. Print.

Glotfelty, Cheryll. "Introduction: Literary Studies in an Age of Environmental Crisis." *The Ecocriticism Reader: !Landmarks in Literary Ecology*. Eds. Cheryll Glotfelty and Harold Fromm. Athens and London: University of Georgia P, 1996.xv–xxxvii. Print.

Grusin, Richard. "Anthropocene Feminism: An Experiment in Collaborative Theorizing." *Anthropocene Feminism*. Ed. Richard Grusin. Minneapolis: Minnesota UP, 2017.vii–xix. Print.

Hall, Jordan. "A Brief History of Human Extinction." 2018. Theatrical script.

Hann, Rachel. "Justice Scenographics: Preparing for Civilization Change in a Time of 'Anywheres' and 'Somewheres'." *Spatial justice 2.0*. Dorchester: Design Studio for Social Intervention, 2020.18–23. <www.ds4si.org/writings/spatialjustice-zine>.

Haraway, Donna J. *Staying with the Trouble. Making Kin in the Chthulucene*. Durham, NC: Duke UP, 2016. Print.

Heim, Wallace. "Performance and Ecology: A Reader's Guide." *Performing Nature: Explorations in Ecology and the Arts*. Gabriella Giannachi and Nigel Stewart. Eds. Oxford: Peter Lang, 2005.405–15. Print.

Heim, Wallace and Eleanor Margolies, eds. *Landing Stages: Selections from the Ashden Directory of Environment and Performance 2000–2014*. London: Crinkle Crankle Press, 2014. Print.

Heise, Ursula K. "Globality, Difference, and the International Turn in Ecocriticism." *PMLA: Publications of the Modern Language Association of America* 128.3 (2013): 636–43. Print.

———. *Imagining Extinction: The Cultural Meanings of Endangered Species*. Chicago: University of Chicago P, 2016.

———. "Introduction: Planet, Species, Justice – and the Stories We Tell about Them." *The Routledge Companion to the Environmental Humanities*. Eds. Ursula K. Heise, Jon Christensen, and Michelle Niemann. London and New York: Routledge, 2017.1–10. Print.

———. "Lost Dogs, Last Birds, and Listed Species: Cultures of Extinction." *Configurations* 18.1–2 (Winter 2010): 49–72. Print.

———. *Sense of Place and Sense of Planet: The Environmental Imagination of the Global*. New York: Oxford UP, 2008. Print.

———. "The Hitchhiker's Guide to Ecocriticism." *PMLA: Publications of the Modern Language Association of America* 121.2 (2006): 503–16. Print.

"Infinitely Yours." By Miwa Matreyek, directed by Miwa Matreyek, performance by Miwa Matreyek. Sundance Film Festival, 27 Jan. 2020, Los Angeles, CA.

Iovino, Serenella. "Hyperobject Covid-19." Web blog post. *Cambridge University Press*. 14 May 2020. Web. 10 June 2020. <www.cambridgeblog.org/2020/05/hyperobject-covid-19/>.

———. "Material Ecocriticism: Matter, Text, and Posthuman Ethics." *Literature, Ecology, Ethics: Recent Trends in European Ecocriticism*. Eds. Timo Müller and Michael Sauter. Heidelberg: Winter Verlag, 2012.51–68. Print.

Iovino, Serenella and Serpil Oppermann. "Introduction: Stories Come to Matter." *Material Ecocriticism*. Eds. Serenella Iovino and Serpil Oppermann. Bloomington: Indiana UP, 2014. Print.

———. "Material Ecocriticism: Materiality, Agency, and Models of Narrativity." *Ecozon@: European Journal on Literature and Environment* 3 (2012): 75–91. Web. 20 Feb. 2014.

Kershaw, Baz. *Theatre Ecology: Environments and Performance Events*. New York: Cambridge UP, 2007. Print.

Latour, Bruno. "Agency at the Time of the Anthropocene." *New Literary History* 45.1 (2014): 1–18. Print.

_____. "An Attempt at a 'Compositionist Manifesto'." *New Literary History* 41 (2010): 471–90. Print.

———. "Facing Gaia: Six Lectures on the Political Theology of Nature." Gifford Lectures on Natural Religion. Edinburgh. 18–28 Feb. 2013. Web. 16 Sept. 2015. <https://macaulay.cuny.edu/eportfolios/wakefield15/files/2015/01/LATOUR-GIFFORD-SIX-LECTURES_1.pdf>.

———. "Is this a Dress Rehearsal?" *Critical Inquiry* 47.S2 (Winter 2021): S25–S27. <www.journals.uchicago.edu/doi/full/10.1086/711428>.

———. "On Actor-Network Theory. A Few Clarifications Plus More Than a Few Complications." *Soziale Welt* 47.4 (1996): 369–81. Print.

Lavery, Carl. "Introduction: Performance and Ecology – What Can Theatre Do?" *Green Letters: Studies in Ecocriticism* 20.3 (2016): 229–36. Print.

———. "Introduction: Performance and Ecology – What Can Theatre Do?" *Performance and Ecology: What Can Theatre Do?* Ed. Carl Lavery. Abingdon: Routledge, 2018.1–8. Print.

Lavery, Carl and Clare Finburgh, eds. *Rethinking the Theatre of the Absurd: Ecology, the Environment and the Greening of the Modern Stage*. London: Bloomsbury Publishing, 2015. Print.

Lazare, Sarah. "'Colonizing the Atmosphere': How Rich, Western Nations Drive the Climate Crisis." *In These Times*. In These Times, 14 Sep. 2020. Web. 02 May 2021. <https://inthesetimes.com/article/climate-change-wealthy-western-nations-global-north-south-fires-west>.

Lewis, E.M. *Song of Extinction*. New York: Samuel French, 2010. Print.

May, Theresa J. "Beyond Bambi: Toward a Dangerous Ecocriticism in Theatre Studies." *Theatre Topics* 17.2 (Sept. 2007): 95–110. Print.

———. *Earth Matters on Stage: Ecology and Environment in American Theater*. New York: Routledge, 2020. Print.

———. "Greening the Theater: Taking Ecocriticism from Page to Stage." *Journal of Interdisciplinary Studies* 7.1 (Fall 2005): 84–103. Print.

Morton, Timothy. "Art in the Age of Asymmetry: Hegel, Objects, Aesthetics." *Evental Aesthetics* 1.1 (2012): 121–42. Print.

———. "Dark Ecology: Philosophy in the Anthropocene." Humanities Without Boundaries. UW-Madison Center for the Humanities, Madison. 19 Oct 2012. Lecture.

———. *Ecology without Nature: Rethinking Environmental Aesthetics*. Cambridge: Harvard UP, 2007. Print.

———. *Hyperobjects: Philosophy and Ecology after the End of the World*. Minneapolis: University of Minnesota P, 2013. Print.

———. "Materialism Expanded and Remixed." *New Materialisms*, Johns Hopkins University, Apr. 13–14 2010: (1–17). Scribd. 04 July 2015. <http://dm.ncl.ac.uk/courseblog/files/2011/03/25830212-Materialism-Expanded-and- Remixed.pdf>.

———. "OOO and Quantum Theory." Web blog post. *Ecology Without Nature*. Timothy Morton, 27 July 2013. Web. 02 Nov. 2016. <http://ecologywithoutnature.blogspot.com.au/2013/07/ooo-and-quantum- theory.html>.

———. *The Ecological Thought*. Cambridge: Harvard UP, 2010. Print.

———. "Zero Landscapes in the Time of Hyperobjects." *Graz Architectural* 7 (2011): 78–87. Web. 12 Feb. 2014.

Murphy, Patrick D. "Grounding Anotherness and Answerability through Allonational Ecoliterature Formations." *Nature in Literary and Cultural Studies: Transatlantic Conversations on Ecocriticism*. Eds. C. Gersdorf Catrin and S. Mayer. Amsterdam: Rodopi, 2006.417–34. Print.

Nixon, Rob. *Slow Violence and the Environmentalism of the Poor*. London: Harvard UP, 2011. Print.

Oppermann, Serpil. "Transnationalization of Ecocriticism." *Anglia: Journal of English Philology* 130.3 (Spring 2012): 401–19. Print.

Phillips, Dana and Heather I. Sullivan. "Material Ecocriticism: Dirt, Waste, Bodies, Food, and Other Matter." *Material Ecocriticism*. Eds. Heather Sullivan and Dana Phillips. Spec. issue of *ISLE* 19.3 (Summer 2012): 445–7. Print.

Read, Alan. *Theatre, Intimacy & Engagement: The Last Human Venue*. London: Palgrave Macmillan, 2009. Print.

Richmond, Joy D. "Introduction: Ecocriticism." *Dramatic Theory and Criticism* 20.2 (2006): 123–6. <https://journals.ku.edu/jdtc/issue/view/284>.

Sewell, Stephen. *It Just Stopped/Myth, Propaganda and Disaster in Nazi Germany and Contemporary America*. Sydney: Currency P, 2007. Print.

Shepherd-Barr, Kristen E. "An Interview with Kirsten Shepherd-Barr, Author of Theatre and Evolution from Ibsen to Beckett." Web blog post. Interview by Columbia UP. *Columbia UP Blog*. 6 May 2015. Web. 20 June 2016. <www.cupblog.org/?p=16543>.

———. *Theatre and Evolution from Ibsen to Beckett*. New York and Chichester: Columbia UP, 2015. Print.

Slovic, Scott. "Editor's Note." *Interdisciplinary Studies in Literature and Environment* 19.4 (Autumn 2012): 619–21. Print.

———. "The Third Wave Ecocriticism: North American Reflections on the Current Phase of the Discipline." *Ecozon@: European Journal on Literature and Environment* 1.1 (2010): 4–10 Web. 12 Oct. 2013.

Smith, Bradon. "Staging Climate Change: The Last Ten Years." *Tipping Points Working Paper.* TippingPoint, 2013. Web. 06 Feb. 2014. <www.tippingpoint.org.uk/wp-content/uploads/2013/08/Staging-climate- change-the-last-ten-years.pdf>.

———. "Love, Loss and Rethinking our Place in the World: Climate Change in Performance." *Landing Stages: Selections from the Ashden Directory of Environment and Performance 2000–2014.* Eds. Wallace Heim and Eleanor Margolies. London: Crinkle Crankle Press, 2014.9–12. Print.

Stevens, Lara, Peta Tait, and Denise Varney, eds. *Feminist Ecologies: Changing Environments in the Anthropocene.* Switzerland: Palgrave Macmillan, 2018. Print.

Szerszynski, Bronislaw, Wallace Heim, and Claire Waterton. *Nature Performed: Environment, Culture and Performance.* Oxford: Blackwell Publishers, 2003. Print.

"This World Made Itself." By Miwa Matreyek, directed by Miwa Matreyek, performance by Miwa Matreyek, Wexner Center for the Arts, 31 Oct. 2013, Columbus, OH.

Wilmer, Stephen Elliot and Karen Vedel. "Introduction." *Theatre and the Anthropocene* 32.1 (2020): 1–5. <https://tidsskrift.dk/nts/issue/view/8763>.

Woynarski, Lisa. *Ecodramaturgies: Theatre, Performance and Climate Change.* Switzerland AG Palgrave Macmillan, 2020. Print.

2 Setting the stage for the material turn and agential bodies

Caryl Churchill's *The Skriker* and *Far Away*

While extending ecocritical thinking to the theatrical medium, which has long been obsessed with characters or personae and their dialogue and with the personal and local scale, British playwright Caryl Churchill grapples with ecological issues in imaginary worlds. In its contemporary (postdramatic) performance genres and orientation towards a non-anthropocentric stage, which has evolved since the 1980s, such theatre takes place in "a community of expressive presences," to use David Abram's expression (*Becoming* 173), and thereby expands our ecological consciousness. Churchill's body of dramatic work experiments with a range of ecological forms and contributes effectively to that expansion, prompting us to conceptualise the ecological alignment of contemporary theatre with all life forms in new materialist terms. Her plays *The Ants* (1962), *Not Enough Oxygen* (1971), *Fen* (1983), *Lives of the Great Poisoners* (1991), *The Skriker* (1994), *Far Away* (2000), *A Number* (2003), *We Turned on the Light* (2006), *Love and Information* (2012), and *Escaped Alone* (2016) are each a landmark in the history of ecological theatre. They contain a great deal of environmental imagery, envisioning a kind of theatre that unites the human with the nonhuman in the face of ecological catastrophe. Under the influence of the political and economic system of neoliberalism, Britain has witnessed an increase in environmental degradation in the differing forms of climatic instability, water pollution, and toxic waste (Campbell). For example, according to John Campbell, it was around 1988 that environmental concerns suddenly acquired a new dimension with the discovery of global warming, caused by increasing emissions of carbon dioxide, methane, and chlorofluorocarbons (CFCs) into the earth's atmosphere (642). Framed within the context of a world dominated by neoliberal globalization, Churchill's plays represent how late consumer capitalism has led to the disintegration of society, the increasing privatisation of social life, the aggressive pursuit for self-interest, as well as the exploitation of natural resources. Theatre in the United Kingdom has embraced these challenges by developing a steady stream of climate-themed plays and performances over the last decade. Among the most prominent are Steve Waters's *The Contingency Plan* (2009), Nick Payne's *If There Is I Haven't Found It Yet* (2009), Zoë Svendsen's *3rd Ring Out: Rehearsing the Future* (2010), Duncan MacMillan's *Lungs* (2011), and Lucy Kirkwood's *The Children* (2016).

DOI: 10.4324/9781003048749-2

Churchill's early works set the tone for her employment of ecological tropes in the rest of her oeuvre, and in her plays of the late twentieth and early twenty-first centuries in particular. For example, the radio drama *Not . . . Enough Oxygen* (1971) is one of the earliest works in which Churchill engages with environmental exploitation. Set in a dystopian future in the year 2010, the play presents a dark vision of a polluted England both thematically and formally, through breathless dialogue, repetitive speech, and gasping. Almost everyone is hiding in tower blocks for fear of the outdoors, which are filled with thick air and acrid smog. Oxygen is scarce, and the character Vivian, who speaks in gasps, is forced to spray oxygen around her room so that she can breathe. Almost all animals and birds have disappeared, and water and plants are in short supply. By depicting such ecological disaster in an urban dystopia, Churchill reveals how consumer capitalism controls the distribution of environmental resources and human access to them (Rabillard 90). For example, in her characteristically experimental style, Churchill writes: "Plants plants would take money. Earth plants earth would all have to come in from the park and the park park authority wouldn't permit. Because hardly any park hardly any park left" (Churchill, *Oxygen* 39). Both parks and the plants in them are hard to come by, and both are constrained by capitalistic exchange value.

In a further instance, in *Lives of the Great Poisoners* (1991), Churchill explores a different form of concern with global environmental risk. The play, called an "eco-opera" (Kritzer, "Systemic Poisons" 164), includes dance and music, and features a group of legendary poisoners including Medea, Dr Crippen, and Madame de Brinvilliers. Ironically, the greatest of them was Thomas Midgley, "an American industrial chemist in the early part of this century," who "put the lead in petrol and CFCs in fridges, two inventions that seemed a good idea at the time but were inadvertently poisonous" (Churchill, *Poisoners* ix–x). Although Midgley was only trying to keep food safe and people healthy, his invention of chlorofluorocarbons caused environmental destruction and pollution. It seems that, as his character says, "[t]hings don't always turn out the way you think. Impossibly horrible things sometimes happen" (38). In this case, the play's ecological concerns anticipate the Anthropocene epoch in which past human activities induce irreparable changes to the global atmosphere and ecosystem. As a further example of this eco-global crisis, Midgley talks about his plan to enhance the growth of crops by intervening in the ozone layer of the earth's atmosphere (66). Churchill's point is that the outcomes of these interventions go far beyond our power and imagination, so that they can no longer be controlled by political and technological practices.

We Turned on the Light, another example of climate change theatre, is a libretto written by Churchill for a choral work composed by Orlando Gough for the BBC Proms in 2006. It addresses consumerism and the depletion of natural resources, and the consequences these may have for the well-being of future generations. Taking up an image Churchill has already used in *The Skriker* (1994), the libretto represents that play's great-great-granddaughter who suffers the effects of climate change while speaking back through time to ask the present-day protagonist: "Didn't you love me?" – to which the protagonist's ghost replies, "Not enough. It's hard to love people far away in time" (Churchill, *We Turned on the Light* n. pag.). The libretto shows how the unknown outcomes of human activities around the world exceed the boundaries of time and place, and increase the vulnerability of humans to environmental change. This global expansion is most evident in *The Skriker* and *Far Away*, as these plays sketch the crossing of material boundaries and respond both thematically and formally to the multi-scalar temporal and spatial variability of the Anthropocene epoch.

Furthermore, in *Love and Information* (2012), Churchill dramatises alternative views about the future of the environment. In the scene "Climate," for example, a woman expresses her fear of the impact of global ecological degradation on future generations while her mother happily trims a Christmas tree. The woman states: "I'm frightened for the children. . . . It's whether they drown or starve or get killed in the fight for water" (54). Concern with ecological uncertainty is further demonstrated in Churchill's latest play, *Escaped Alone* (2016), which artfully opens up contemporary eco-dystopias. The play juxtaposes conversations among four women in their seventies as they sit in a back garden and chat about a range of mundane and trivial subjects, on the one hand, with a series of monologues delivered by one of them, on the other. The monologues represent a different world from the women in the garden by recounting global catastrophes and the disintegration of modern life in the wake of environmental devastation. The women's light-hearted gossip, mundane comfort, and idyllic garden contrast with the surreal picture presented on stage, which extends well beyond their imaginations. The benign pleasures the women are currently enjoying are interspersed with planetary meltdown.

In the rest of this chapter, I read the plays *The Skriker* (1994) and *Far Away* (2000) from a material ecocritical perspective to explore how these plays, through their experimental staging and language, are engaged materially and theatrically in redefining human agency and representing environmental degradation. Although they approach the Anthropocene in an implicit rather than an explicit way, these plays are innovative. They are proper examples of the theatre of the Anthropocene that explore and expand vocabularies, epistemologies, and ontologies related to global ecological questions. Churchill's dark imagining of global-scale ecological assemblages in *Far Away*, and of hazardous landscapes and poisoned bodies in *The Skriker*, which Elin Diamond calls an "ecological parable" ("Tragic Materialism" 754), have implications for our fundamental aesthetic, ethical, and metaphysical thinking in the Anthropocene epoch. The plays' discontinuous forms exemplify Una Chaudhuri's idea of a new dramatic strategy that is no longer built upon "the methods of humanist inquiry" and "individual experience" (Chaudhuri, "Anthropo-Scenes" 312) but, rather, on the disruption of "the imagination" and "the smooth functioning of every register or convention of performance" (305). The current chapter builds on this approach, inviting us to rethink our vision of human agency on broader scales of time and space and to comprehend the material forces that act all around us and within us, far beyond our habits of sensory and cognitive apprehension.

1 Trans-corporeal materialities in the theatrical space of *The Skriker*

The Skriker was first performed in the Cottesloe auditorium of the Royal National Theatre, London, in January 1994. By moving between the human world and the underworld, the play recounts the story of humans in relation to the spirit world of the Skriker, the eponymous ancient fairy. The human world of the play is circumscribed to two young women: Josie, who murdered her newborn baby when suffering from mental illness, and Lily, who is pregnant and homeless. The other characters in the play are supernatural creatures, nymphs, spirits, fairies, and goblins from fairy tales, and all are led by the Skriker. The Skriker, who is described at the very beginning as "*a shapeshifter and death portent, ancient and damaged*" (Churchill, *The Skriker* 1), attempts to draw the two impoverished young women into its seductive underworld. In their attempt to make a life for themselves, Lily and Josie are haunted, tormented, and lured on by the Skriker's different guises and identities throughout the play.

Given that it takes place in a fractured world amidst a whirlwind of harmful environmental forces, *The Skriker* has more recently been understood as sounding an urgent warning about the danger of global ecological damage through eco-fairy tales. Churchill has often enthralled her audiences with mythological figures and motifs, fairy tales, and folklore in plays such as *A Mouthful of Birds*, *A Mad Forest*, and *Fen*. Like the Vampire in *A Mad Forest*, the figure of the Skriker is a mutative, menacing, and dark force. However, *The Skriker* is distinctive because its folkloric figures contribute to the instability of time and space in a contemporary high-risk environmental context and, more importantly, are materialised in dramatic action, rather than simply populating the margin of discourse or appearing in the characters' narration, as happens in *Fen*. The mythical figures in *The Skriker* are mischievous, fairy-like spirits derived from English, Scottish, and Irish folklore; they have a strong connection to the environment and are rooted in the natural elements of earth, air, fire, and water (Bane). According to Amelia Howe Kritzer, these figures "bring associations with the inherent dangers of nature lurking below its surface – in a rippling river . . . that can sweep a human away and drown them in an instant, or an enticing forest with the potential to . . . trap the inexperienced explorer" ("Damaged Myth" 115). For instance, Jenny Greenteeth is a Lancashire nymph or water-fairy who chases children into ponds; Rawheadandbloodybones is also a Lancashire water demon who drags children into ponds and devours them; Kelpie, an Irish and Scottish figure, haunts rivers and appears as a water horse and body eater; and Black Annis, a bogeyman from English folklore, lives in a cave in a forest oak and hounds children. As is evident, these folkloric figures are also preoccupied with causing harm to children, stealing them and sucking their blood. The Skriker's name itself derives from "a Lancashire term for 'a shrieker, a screamer'" (Wolf n. pag.) who announces impending disaster and death. Kritzer writes that "[t]he Skriker's name, an archaic form of the word *screamer*, . . . joins an emblem of fatality to the concept of terrifying and unavoidable consequences of previous action" ("Damaged Myth" 114). The Skriker appears in other myths and legends as well. For example, it is mentioned in the *Encyclopedia of Fairies in World Folklore and Mythology*: "Originating from the fairy lore of Yorkshire England, Skriker is a fairy animal appearing as a Black Dog or Barguest; it also has the ability to become invisible" and can also signify a death omen (Bane 308). Furthermore, in some folk tales Skriker is given an alternative name, "Trash," that accords well with its association with ecological destruction and the polluted planet in the play (Briggs 222). Drawing on the history of folkloric figures, Churchill chooses mythical characters to depict a damaged, sick, and poisoned world of revenge and fury.

Through her use of history, Churchill explores the present and future of our planet. Her dark depiction of ancient mythological characters heightens the intensity of the play's warning about the destructive processes taking place in the material world while also communicating the pervasive nature of the negative forces that cause them. For this reason, Geraldine Cousin argues that the spirits offer "a manifestation of human pollution of the planet" (95). More precisely, the figures, by materialising the damaged side of the play's landscape, appear as malignant and disturbing forces, each telling its own story of what has led us to the depth of ecological catastrophe. The playwright's description of the Skriker as "*a shapeshifter and death portent, ancient and damaged*" (Churchill, *The Skriker* 1), hundreds of years old, suggests the inevitable end of the ancient natural world and the emergence of a world in environmental crisis. The history of industrialisation has produced environmental pollution that now takes the form of a sick, revengeful, and raging Skriker, whose shriek announces a shift from fairytale to nightmare. The Skriker recounts a natural fairyland where humans still respected the earth, but that woodland fantasy is far from the ominous reality in which the play is set, where the world appears out of balance. For example, the Skriker calls out human indifference to and lack of respect for the natural (fairy) world in its introductory monologue, attacking the modern world and its destructive tendencies: "They used to leave cream in a sorcerer's apprentice. Gave the brownie a pair of trousers to wear have you gone? Now they hate us and hurt hurtle faster and master. They poison me in my rivers of blood poisoning makes me arms swelter" (4). While humans used to appease nature spirits and thank them for their services, now "they" (humans) poison them in their "rivers of blood" and abuse the land. The phrase "hurtle faster" evokes humankind moving faster and faster as a consequence of progress, yet the playwright draws attention to the fact that "hurtle faster" also contains the word "hurt," suggesting that such progress comes at the cost of hurting the natural environment.

Elsewhere the Skriker accuses "you," or humans as a species, lamenting the dark reality of environmental contamination from anthropogenic activities in a more obvious way: "how you make poisons? . . . You people are killing me, do you know that? I am sick. . . . Just so long as you know I'm dying, I hope that satisfies you to know I'm in pain" (15). Throughout the play, the Skriker illustrates the ways in which humans have made the earth sick and spoiled its natural habitat, with references to human-induced global change that has damaged the ozone layer and produced "toxic waste" and "poison in the food chain saw massacre" (31–2). The Skriker appears in a hotel bar in the guise of an American woman who wants to know about human-made satellites and their strangely ubiquitous presence. The woman asks one of the two human characters, Lily, questions about how the television image is broadcast and how aircraft work – how "it's happening *there* and it's *here*" (14). She is referring to the contraction of space in time, a hallmark of modern industrialised societies. Lily, however, cannot provide a coherent and adequate explanation. Her failure to comprehend the world of technology indicates that the impact of human actions reaches far beyond our temporal and spatial dimensions, and their ongoing consequences reach far beyond our knowledge and imagination.

To all these human mishaps, the Skriker's response is: "We'll follow you on the dark road at nightingale blowing. . . . Revengeance is gold mine, sweet" (4–5). The Skriker seeks vengeance on the play's human characters, Josie and Lily, who serve as exemplary participants in a modern world driven towards progress. The Skriker's desire for retaliation is accompanied by its vulnerability and need for attention and love. The Skriker "can't tolerate being disliked" (44) and, in its different transformations, demands love from the two human characters. For instance, on one occasion, the Skriker's request for love and attention is denied, which makes Lily and Josie sick and causes them to vomit. To punish the two women for their complicity in this devastation, the Skriker appears as a derelict woman and makes live toads emerge from Josie's mouth. On another occasion, coins spit from Lily's mouth when she tries to speak. These scenes, in which the women appear contaminated and machine-like, imply wider scales of pollution, or what the Skriker refers to as "the waste ground" (47). Hazardous waste and toxic substances produced by human activities and rampant consumption have permeated and poisoned the land, rivers, air, soil, minerals, cells, bacteria, molecules, and atoms, which are now revealed in the form of the enraged and sick Skriker and other wounded fairy figures of the underworld. As the Skriker warns, poisoning makes its arms "swelter," which evokes the growing impact of environmental change – for example, the sweltering weather caused by global warming. The more the Skriker is ignored and disliked, the more it "swells" with a stronger ravaging power.

The Skriker, in the guise of a thirty-year-old man, announces an apocalyptic world of prevailing meteorological extremes and unprecedented catastrophes – a world that is indeed emerging from the Anthropocene, the geological age of humans. The following lines further highlight this concern:

> Earthquakes. Volcanoes. Drought. Apocalyptic meteorological phenomena. The increase of sickness. It was always possible to think whatever your personal problem, there's always nature. Spring will return even if it's

> without me. Nobody loves me but at least it's a sunny day. This has been a comfort to people as long as they've existed. But it's not available any more. Sorry. Nobody loves me and the sun's going to kill me. Spring will return and nothing will grow. Some people might feel concerned about that. But it makes me feel important. I'm going to be around when the world as we know it ends. I'm going to witness unprecedented catastrophe. (43–4)

The Skriker's story of the real world is not a fairytale ideal of a sustainable environment located somewhere out there, but a scenario of post-anthropocentric environmentalism in which the nonhuman world has become a dynamically agential force with unpredictable impact, causing unprecedented displacements. Churchill's portrayal of this impending ecological disaster cautions us that, just as the modern world and its destructive potential have displaced and damaged the ecological system, the earth – through nonhuman figures – acts with its own agential power, breaching the very boundaries of the human world as such.

The dramatic world of *The Skriker* is also governed by an assemblage of shifting and wordless creatures. With their distinctive appearances, arrangements, gestures, and mutative identities – and persistent and haunting presence – these creatures are distinguishable from, and yet invisibly penetrate, the human world (see fig. 2.1 and 2.2). Churchill shapes this structure according to a non-anthropocentric set of principles. Alongside Josie and Lily's interactions with the Skriker, the presence on the stage of mute and ever-present nonhuman

Figure 2.1 The fairy creatures. *The Skriker.* The Wesleyan Theater Department, Middletown, CT, 2009. Directed by Bob Bresnick. Photo: Olivia Bartlett Drake.

Figure 2.2 The fairy creatures. *The Skriker*. The Tulane Department of Theatre and Dance, 2011. Directed by Rebecca Frank. Scenic and costume design by Catherine Jacot-Guillarmod. Photo by Rebecca Frank and Catherine Jacot-Guillarmod.

characters, who dance at the edges of the action and play a background role, adds layers to the narrative and thus decentralises the human figures. For example, this collection of surreal and somber images begins with Johnny Squarefoot, "*a giant riding on a piglike man, throwing stones*" (1). This strange opening gesture of a giant riding on a pig – or rather, a ghost with the body of a man and the head of a pig that is associated with land or earth and haunts the Isle of Man – prefigures the Skriker's shrieking arrival onstage. The giant's act of throwing stones also suggests his rage towards his enemies – the human beings who have damaged the authentic and sustainable older fairy world. In the scene following the Skriker's opening monologue, the Kelpie, "*part young man, part horse*" (5), watches over Josie's bed in a mental hospital, though the girls do not notice him. In folklore, Kelpies assume human form to leap onto horses and kill travellers, or they approach unsuspecting young women as lovers before leading them away and devouring them underwater. Kelpies are also described as shape-shifting water horses with the power to command water and drown their victims, especially wandering children or unwary young men (Rose 178). It is also said that Kelpies prevent watermills from turning at night and are capable of destroying them (178). In Churchill's play, these creatures appear to hate humans for destroying their home and polluting the water. In a later scene, when the now-pregnant Lily is at a hotel bar talking to the Skriker in the shape of an American woman of about forty, Spriggan – grotesquely ugly, ten feet tall, and invisible to Lily – arrives. According to folk tales, this malicious fairy is found on land and in the water and is thought to be the source of misfortunes such as spoiled crops, severe weather, and illness. Spriggans are also believed to steal children and replace them with changelings. The fairy's presence in the bar suggests that he is waiting for Lily's baby to be born so that he can snatch it. Each part of the play is accompanied by the corporeal presence of nonhuman figures who embody warnings that are implicit in their stories, though the audience does not hear directly from all of these characters. These warnings principally concern the human exploitation of land and natural resources.

The fairy figures, along with other mute creatures throughout the play, constitute examples of silent violence emerging just beyond our grasp and understanding. This violence unfolds just as their natural habitats – air, water, or land – are poisoned by human activities. For example, rivers and lakes are polluted by radiation, and lands are contaminated with toxic wastes left by industry. By appearing and disappearing, the goblins – each of whom illustrates an uncanny narrative of degradation – represent the threatening nonhuman forces that are potentially mobilised in the context of ecological degradation; at the same time, these forces permeate every moment of the human characters' lives. Such dark forces resonate with Timothy Morton's concept of "strange strangers." The ecological intimacy of the global network of interconnectedness that Morton calls the "mesh" is not complete in itself and is often accompanied by strangeness, darkness, gaps, and absences. Morton proposes the notion of strange strangers as an alternative to that of nonhumans and animals. This concept considers "the limit of our imagining," signaling that each entity in the mesh is always defined in relation to unpredictable forces, so the more we know about them, "the more we recognize our connection with them and the stranger they become" (Morton, *Ecological Thought* 17). The coexistence of strangeness and intimacy in the decentered system of mesh calls for non-anthropocentric narratives and the need to attend to nonhumans or strange strangers, who persist right "under our skin" (Morton, *Ecological Thought* 119). Morton's concept of strange strangers illuminates the onstage omnipresence of the unknown and hostile fairy figures at work in the backdrop of every scene in that they become recessively stranger as they come closer into contact with human characters, Lily and Josie, who fail to recognise them. The more entangled these material forces are with the girls' bodies and lives, the more ambiguous, stranger, and more unpredictable they become.

Thrumpins and Blue Men also obviously resonate with Morton's concept of "strange strangers" because they surround humans and pervade their everyday lives though people cannot perceive them. For example, without being visible to Lily and Josie, chattering Thrumpins or evil guardians ride on the backs of a group of unknowing businessmen who have gathered for a meeting but cannot hear each other. Thrumpins have the power of taking the lives of the businessmen by whispering words of death into their ears. Also, a picnicking family is crowded by a group of Blue Men or sea spirits. Spirit bodies haunt the beach while swimming alongside ships in their efforts to drown sailors. These scenes present a series of theatrical images that signify the invasion of unseen beings, and their interaction with human forces explodes boundaries in the material world. Like strange strangers, these spirits and their imperceptible and threatening presence are a reminder that human mastery to shape and change the world is restricted, evoking the uncertainties and complexities of worlding practices.

In *The Skriker*'s production at the Royal Exhibition Theatre as part of the Manchester International Festival in July 2015, critics acknowledged that its environmental themes were even more timely and disturbing than they had been two decades earlier when the play premièred at the National Theatre in London. Reviewing the production in the *Guardian*, Amy C. Chambers remarks that this revival is, for director Sarah Frankcom and actor Maxine Peake, who played the shape-shifting Skriker, "an environmentalist call to action" that responds "to conversations in the media concerning capitalism and its impact on the natural world" (n. pag.). For Peake, the Skriker embodies "*our world in constant environmental crisis*" and, more broadly, "*the Earth's last cry for help*" (Chambers, emphasis in original, n. pag.). This perspective retains its environmental relevance and situates the play at the intersection of science and art by "bringing science issues to the stage" (Chambers n. pag.). The production was also influenced by Naomi Klein's *This Changes Everything: Capitalism vs. the Climate* (2014), which the ensemble cast read in preparation for the performance. The book, which Peake refers to as "a bible piece," addresses the most crucial threat humanity has ever faced: the tensions among our economic system, individual and social health, and the natural world (Chambers n. pag.). This production turned the play into an experience of being inside a fractured world and responded to the question of how we should tell stories in an age of environmental challenges.

Temporal-spatial variability in the ecological context

The aforementioned examples of the fairy figures' dynamic presence illuminate their relation to spatio-temporal dislocation, in which the boundaries between the real world and the underworld, the everyday and fantasy, are dismantled. The instability of time and space, which Diamond describes as "*temporal* distortion" (*Unmaking Mimesis* 100) and Candice Amich as "time-space compression" (394), is caused by the collision of the fairy creatures' ancient and contemporary worlds. For Diamond, retelling fairy tales in the context of contemporary global capitalism demonstrates a break with past traditions ("Feeling Global" 481–2). Building on Diamond's reading, Amich emphasises that the fairy figures' manipulation of the flow of time and space symbolises the fluidity of late capitalism. My analysis extends Amich's focus on the play's exploration of "the ravages and intoxications of globalization" (394), recognising how a much broader network of corporeal interchanges – simultaneously economic, political, cultural, scientific, and material – unsettle our normative ideas of time and space. The play shows the distortion of spatio-temporal logics in the context of these corporeal interchanges by insisting on the fluid boundaries between the material body and the physical environment and evoking a non-anthropocentric world of materiality and performativity while engaging different spaces, temporalities, bodily transformations, and a fractured form of language. I am thus inclined to agree with Diamond's argument that *The Skriker* enacts a posthuman displacement of tragedy from the human onto nonhuman life ("Tragic Materialism"). In particular, this approach contributes to my analysis of how the play's dramatic worlds of temporal and spatial discontinuity are pertinent to the posthuman trans-corporeality of material bodies. According to Stacy Alaimo, "[t]rans-corporeality is a posthumanist mode of new materialism," and it means that "all creatures, as embodied beings, are intermeshed with the dynamic, material world, which crosses through them, transforms them, and is transformed by them" ("Trans-corporeality" 435). I argue that the idea of trans-corporeality helps us understand the play's ecocritical concerns. To delineate the material conditions of a non-anthropocentric dramaturgy in the play, this concept allows us to envision a fluid approach to temporal and spatial boundaries by interpreting and identifying bodily changes or bodily transgressions unraveled under the influence of environmental changes and across time and space.

In its refusal to occupy one position, the Skriker crisscrosses between the underworld and multiple places in the real world, while its unexpected entrances and mutations fragment the narrative and disrupt temporal and spatial logics. For example, in the scene following the Skriker's opening monologue in the underworld, the figure embeds itself in the invisible workings of the hospital. When Lily goes there to visit Josie, who has been institutionalised for killing her child, Josie tells her friend that she is haunted by someone she thought was a patient, but who is actually hundreds of years old. When Lily leaves the stage, the Skriker enters as a woman of about fifty: "*Dowdy, cardigan, could be a patient. It is the* SKRIKER" (Churchill, *The Skriker* 9). On another occasion, the Skriker dwells in a sofa and Lily feels its presence as an icy, freezing cold sensation that comes upon her. Immediately after that, the Skriker leaps out of the sofa and changes into a fairy and then into a small child in a park. Because of such distortion of boundaries and overlapping scenes, the play's real world does not seem any more predictable than the underworld with which it begins. Churchill seeks to disturb what Alaimo calls the "comforting, conventional sense that the [environmental] problem is out there, distinct from one's self" ("New Materialisms" 561). Both the real world and the netherworld are shattered, uncertain, and unknowable, unsettling optimistic narratives of environmental sustainability. The enraged Skriker with its damaged body persuades us that while it has stepped out of the ancient natural world, its current situation is dependent on human decisions in the present.

Such acts of transmutation – across time and space – displace the human characters from their centrality and extinguish their capacity to assign static meaning to the world. This trans-corporeal and mutative environment asks audiences to imagine the two young female characters and themselves in what Alaimo describes as a "constant interchange with the environment . . . that allows for both the unpredictable becomings of other creatures and the limits of human knowledge" (*Bodily Natures* 22). In this situation, trans-corporeality is not an affirmation of human exceptionalism as well as ecological harmony but includes, as Alaimo asserts, "particular moments of confusion and contestation that occur when individuals and collectives must contend not only with the materiality of their very selves" (17) but also with often unexpected and dangerous material agencies, substances, and landscapes of risk. This multifaceted world of trans-corporeality is demonstrated through the play's experimental staging as well as its spatio-temporal discontinuities. For example, the play's collaborative staging at Manchester's Royal Exhibition Theatre represents the transformative nature of the Skriker as a major formal and thematic element by dismantling the distinction between the actual world and the underworld: audience members sit at long wooden tables that form the stage level for much of the action. In this way, the usual barrier between actor and observer is broken, while audience members themselves become part of the performance and are forced to participate in its damaged story (C. Bennett). The play's flexible space is populated by a dozen or so folkloric figures and a chorus of singers flaunting about convulsively around the audience. The actors move continually in between the audience and the tables. For example, when the Skriker takes Lily and Josie to the underworld and then back to the real world, audience members are only inches away from the tumult of such transitions. Such staging shows how the audience is, in fact, part of the fractured and destructive world that humans have made. Such a transformed auditorium, designed by Lizzie Clachan, provides a dramaturgical space of trans-corporeality that traces the material "interconnections, interchanges, and transits" across human bodies, nonhuman bodies, and the wider material environment (Alaimo, *Bodily Natures* 2).

Along with the play's experimental staging, language is also central to the disruption of temporal and spatial logics. As the following lines show, the Skriker's surreal monologue is full of rhymes, broken and fragmented words, and shifts in verb tense:

> Tell them one thing not to do, thing to rue won't they do it, boo hoo's afraid of the pig bag. Open bluebeard's one bloody chamber maid, eat the one forbidden fruit of the tree top down comes cradle and baby. Don't put your hand in the fountain pen and ink blot your copy catching fishes eyes and gluesniffer. So he puts his hand in and wail whale moby dictated the outcome into the garden maudlin. Everything gone with the window cleaner.
>
> (Churchill, *The Skriker* 3)

It is difficult to say exactly what this exemplary passage is "about," but it is possible to say that the Skriker shrieks its damage and sickness. The unhindered flow of words moves among images of the past, the present, and a dystopian future by linking fairy tales with everyday life. While this perplexing monologue contains several references to literary works and folklore in which human beings suffer in return for the damage they have done to the earth, which is embodied by the Skriker, it also exemplifies spatio-temporal distortion as a prologue that foreshadows events to come. This opening monologue taunts the human world with its recklessness and greed. For example, the Skriker refers to the classic fairy tale of "Bluebeard" in the line "Open bluebeard's one bloody chamber maid, eat the one forbidden fruit" (3), which foreshadows Josie's consumption of forbidden food and poisoned water in the underworld. The Skriker continues with another image of human greed as the language turns violent and relentlessly overrides sequential connections. The words "Don't put your hand in the fountain" (Churchill, *The Skriker* 3) prefigure the Skriker's command to Josie not "to touch the water in the fountain" (33), as it is dangerous and ends in blinding and death. Josie, however, does not heed the warning and "*goes to the fountain and almost puts her hand in the water*" (34). The Skriker's shrieks get louder and louder, and its anger is identified with a "wail whale moby" (3) in an allusion to Herman Melville's *Moby-Dick; or, the Whale.* Like Captain Ahab, whose persistent attempts to kill Moby Dick resulted in his own destruction, Josie's obsessive curiosity and eagerness to consume the underworld's deceptive "glamour" take her "further and murder in the dark" (49). Beginning with the nightmarish visions of fairy tales and combining them with the present and future life of two young women, *The Skriker* evokes an "ecological parable" (Diamond, "Tragic Materialism" 754), a dark and absurd image of the end of things.

The Skriker's parodic playback of human language usage in mangled form highlights the play's transformative and mutative themes of trans-corporeal materiality while offering a theatrical model of the risk and terror associated with ongoing changes in the material world. Such an overlapping and fractured language – from the material flexibility of the single word to the repeatedly changing dynamics of the larger groups of words – is a narration of "experimental language" (Amich 395) that expresses language's materiality. An experimental approach to the Skriker's language, as sentences and words move both forwards and backwards at once and between different tenses and historical and cultural references, exemplifies the idea that language parallels the spatio-temporal disruption of ecological harmony. Of course, the dramatisation of this wild and frantic language in performance has a different effect from that of the written text in the play's script. This form of language play – with its incoherence and uncontrollable, unstoppable words – makes a dramatic situation wherein part of the language's possible significance, meaning, and agency is invisible to the audience who watches the play and is not able to read along and discern fully the action and its effect. As a critical praxis of eco-dramaturgy, the Skriker's language challenges the anthropocentric perspective of the world by dismantling mastery in narrative and interpretation or resisting audience access to a pattern of thought, cognitive tendencies, and "a position of interpretive mastery" (Wilson, "Failure and the Limits" 187). If we take the Skriker's refusal to occupy a stable position in its movements and language to be an expression of messy entangled human-nonhuman relations, the play invites us to think about how the audience's loss of interpretive authority indicates the loss of mastery and control over ecological changes – as the complexity and globality of these changes cannot be fully assimilated into human imagination.

Language itself enables us to conceive of trans-corporeality, which renders the environment a mutable and multifaceted space of material agencies and interchanges. For example, the Skriker's description of the physical world emphasises this concern: "wide open wide world hurled hurtling hurting hurt very badly" (Churchill, *The Skriker* 31). The semantic assemblages of "hurled," "hurtling," "hurting," and "hurt" evoke damaging physical struggles. While the Skriker's language embodies landscapes of risk in Churchill's dramatic world, language itself is not distinct from the materiality of that risk. Such an interpretation asks us to see the capacity of language in its negotiation, not only of cultural relations, but also of the materiality of ecological processes, as language undertakes a material and semantic metamorphosis. This perspective is defined as the "agentic entanglement" of world and text and of matter and discourse (Iovino and Oppermann, "Stories Come to Matter" 10). As Serenella Iovino and Serpil Oppermann argue, instead of separating the discursive and the material and "concentrating on texts and seeing how they 'reflect' the world's phenomena," the idea of agentic entanglement "allows us to actively participate in a creative process in which material levels and levels of meanings emerge together" (9–10). In that respect, the Skriker's long introductory speech can be read as a form of "material narrativity," to use Iovino and Oppermann's term, in which the material world and human discourse are entangled. Such narration is created by the play's manipulation of language – materially, semantically, temporally, and spatially – and catalyses new modes of thinking about environmental instability and trans-corporeal understanding of ecological relations, processes of transformation, and agentic forces. Further, the Skriker's foregrounding of the materiality of language offers a sense of how alternative modes of being and understanding might sound, implying that the process of making meaning cannot be limited to what humans think and understand. Therefore, we cannot think of nonhuman nature "as exterior to human language, but instead . . . the process of making meaning is an ongoing one, a process that includes nonhuman nature as a participant rather than as an object of inquiry" (Alaimo, *Bodily Natures* 42). The Skriker's strange, experimental, and sensory language communicates with other forces, substances, meanings, and signs that coexist with our bodies but are beyond our control and prediction, bringing forth a palpable sense of trans-corporeality.

The play's compression and expansion of time is another example of how it promotes trans-corporeal awareness of ecological relations by illuminating the experience of simultaneously being here and there or inside and outside. Ages are compressed into an instant, or a second is extended into eons as Lily and Josie time travel to the underworld. By morphing into various shapes and moving in time and space, the Skriker seeks to lure Josie and Lily into its grotesque underworld. On the one hand, Josie thinks she has been there for a long time when in fact only an instant has passed. On the other hand, Lily, who believes that going to the underworld will only take a second, is confronted by a desperate vision of the future. She finds that she has left her child behind and has moved into the far future, where she meets not only her granddaughter, who is an old woman now, but also her great-great-granddaughter, who is deformed due to the effects of anthropogenic environmental toxicity and pollution.

Such examples of displacement can be reconsidered in terms of trans-corporeality when we recognise that "humans are the very stuff of the material, emergent world" (Alaimo, *Bodily Natures* 20). From this non-anthropocentric perspective, Lily and Josie are thrown into a realm of interconnected agencies at the same time as they encounter a complex mass of nonhuman forces and timelines that transcend any human-centred point of view. As the characters undergo a transgression of temporal and spatial limits, they illustrate, in Alaimo's words, that "trans-corporeality entails a rather disconcerting sense of being immersed within incalculable, interconnected material agencies that erode even our most sophisticated modes of understanding" (17). Lily and Josie come up against a multiplicity of interwoven temporalities in which the past, represented by the fairy figures, is not left behind but rather lasts. The past becomes entangled with ongoing changes in the environment, transmuting into a different narrative, which in turn creates alternative presents and uncertain futures. In discussions of the Anthropocene, the present is juxtaposed with the deep time of fossil fuel formation and the catastrophic future of those whose lives will be deeply affected by our current actions. Because the play situates itself in a geological timescale, trans-corporeal implications ensue. Its human characters are positioned as the victims of a long history of human activities that have ironically resulted in a variety of intricate and unpredictable environmental changes whereby the forces of deep time flow in bodies. In particular, Alaimo's understanding of the material self in relation to the Anthropocene epoch attributes to humans a lack of control over time and place and "a profound shift in subjectivity" (20). The uncertainty emanating from this sort of trans-corporeality permeates the materiality of the characters' very selves, reconfiguring them into posthumans (154).

Toxic bodies: pathways toward posthuman and material ethics

The aforementioned spatio-temporal issues in the play open up possibilities for exploring the corporeal and material interchanges among toxic human and nonhuman bodies in a fragmented eco-global space. The play illustrates how the human body is not a pre-constituted, autonomous, and unchangeable unit. Rather, the boundaries of the human self are transgressed and reterritorialised into toxic bodies by being part of a complex web of interactions. As Lily and Josie enter "a mobile space that acknowledges the often unpredictable and unwanted actions of human bodies, nonhuman creatures, ecological systems, chemical agents, and other actors," the play demonstrates a "hazardous trans-corporeality" that creates the potential for "toxic bodies" (Alaimo, *Bodily*

Natures 20). In this ecological network, as Alaimo points out, "the most palpable trans-corporeal substance is food, since eating transforms plants and animals into human flesh" (12). I show how the play dramatises the power, agency, and mutability of food, as the basis for human existence, in relation to the toxicity of human bodies, fostering a non-anthropocentric ethics of worlding practices.

The Skriker's monologue includes warnings about the risks of eating enchanted and contaminated food, such as "Never eat a fruit or puck luck pluck a flower if you want to get back" (Churchill, *The Skriker* 3) and "A hungered yours hundred years later. And a bit a bite a bitter bread and he was crumbs crumbling to dust panic" (4). As Cousin argues, food appears as "the most disturbing element" in the play because "what should give life represents contamination and danger" (94). In one story, trans-corporeality among various "bodily natures" occurs when Josie drinks and eats contaminated food in the underworld and submits to its deceptive glamour (see fig. 2.3 & 2.4). In her first impression of the Skriker's underworld kingdom, Josie sees it as a grand palace of enchantment with an overwhelming banquet table, although "*some of the food is twigs, leaves, beetles*" (Churchill, *The Skriker* 29). This frightening banquet scene is a turning point both in production and in highlighting the idea of toxic bodies living on a toxic planet. As Josie and the Skriker arrive, the latter in the guise of a fairy queen, the other creatures burst into song to convince

Figure 2.3 Mary Ewald as the Skriker. Mariel Neto as Josie. *The Skriker*. The Erickson Theatre, 2012. Directed by Janice Findley. By Permission of Janice Findley Productions ©2012. Photo by Paul Joseph Brown, Jyh-Lurn Chang, and Julia Salamonik.

Figure 2.4 Mary Ewald as the Skriker. Mariel Neto as Josie eating the contaminated food. *The Skriker.* The Erickson Theatre, 2012. Directed by Janice Findley. By Permission of Janice Findley Productions ©2012. Photo by Paul Joseph Brown, Jyh-Lurn Chang, and Julia Salamonik.

Josie to drink and eat. Then a Hag "*rushes in shrieking . . . seizes food, scattering it, searching*" while saying: "Where's my head? where's my heart? where's my arm? where's my leg? Is that my finger? that's my eye" (29). The Hag, who was dismembered by the spirits, begs for her body parts back: "They cut me up. They boiled me for dinner" (29). Although it is not clear exactly why the other creatures are disgusted by her presence, one possible explanation lies in the Hag's human status. She is not a fairy creature but an old woman and an outcast from society in the real world above, and thus the spirits are appalled by her presence (Meaker). Following this scene of dismemberment, which prefigures "a cannibal feast of human flesh" (Chambers n. pag.), the Lost Girl warns Josie against eating anything: "It's twigs and beetles and a dead body. Don't eat or you'll never get back" (Churchill, *The Skriker* 30). However, Josie finally drinks with the Skriker and happily partakes of the food, failing to distinguish between cake and twigs. The scene leads immediately to confusion, silence, and the disintegration of the feast. Josie is left in a miserable state while she scrubs the floor on her hands and knees. Her sweet body is poisoned by the toxic foods she has absorbed. This dark image evokes a wider sense of trans-corporeality by which potentially dangerous substances and forces move across her body and mark it over a lifetime.

In this respect, the play enacts what Alaimo calls the "traffic in toxins" – that is, the transformations of chemical matter as it moves from production to consumption and the impact these transformations have on a vast chain of human and nonhuman bodies (Alaimo, *Bodily Natures* 18). Such a space of trans-corporeality is full of uncertainty and confusion because, according to Alaimo, it includes "invisible," "unexpected," "dangerous," "imperceptible," and "devilish" material agencies and substances that are in the process of becoming for millennia and across a global network, "especially as they are bound up with social, economic, and political forces" (147). We can see this view of trans-corporeality across deep history in the representation of the fairy banquet table in the underworld, which "is set and indeed has been set for millennia" (Midgley "Food and *The Skriker*" 107). What the Skriker tellingly reveals is that, given the dominance of human greed, banal consumerism, and indifference to environmental change, nothing is more certain than that the world is being gradually poisoned by devilish and dangerous agencies beyond humanistic conceptualisation. The Skriker laments the decline in the taste of human flesh that is like "salmonelephantiasis, blue blood bad blood blue blood bad blood blah blah blah" (31), after it was polluted by this corrupt environment. The word "salmonelephantiasis" evokes salmonella food poisoning and elephantiasis that causes severe swelling in limbs. When Josie asks, "And now no one tastes any good?" the fairy figure complains that no human flesh tastes good anymore: "Dry as dustpans, foul as shitpandemonium. Poison in the food chain saw massacre" (Churchill, *The Skriker* 32). In response to Josie's offer of her blood, the Skriker says bitterly: "Haven't I slipped lipped lapped your pretty twist wrist for years and fears? What's happy new, what's special brew hoo?" (32). "Pretty twist wrist" would seem to be an allusion to the double helix of human DNA, which is affected by environmental pollution. In this sense, the human body gradually mutates into hazardous toxic waste. Of course, not only her blood but also Josie's memories and dreams are dismissed by the Skriker: "You're dry as dead leaves you behind" (33). It is likely that Josie's blood, flesh, memories, and dreams have all been decayed and poisoned by unwanted and unpredictable trans-corporeality, with the Skriker as the physical embodiment of the sick and toxic environment.

In response to the traffic in toxins, the play's narration of trans-corporeality can be further understood by considering its references to a new baby, who emerges as one of various bodies affected by an unwanted entanglement with the wider environment. Throughout the play, the Skriker tries many times to take possession of a human child to make itself stronger. In the interval between Josie's return from the underworld and the birth of Lily's baby, the Skriker, in the guise of a "*smart woman in mid thirties*," tells the audience that Lily has given birth to a baby: "a booby a babbly byebye booboo boohoo hoooooo" (Churchill, *The Skriker* 36). Lily picks up the sentiment in a sad, telling speech: "Everything's shifted round so she's in the middle. I never minded things. But everything dangerous seems it might get her. I know she's just . . . But if she wasn't all right it'd be a waste, wouldn't it" (37). Lily's concern about the future of her child suggests her suspicion of the environment, which is a grotesque wasteland filled with strange and unknown forces and a fragmented flow of time and space. This sense of uncertainty and mistrust is reinforced when Josie thinks "[t]here's something watching us" (38), which is confirmed by the invisible presence of Rawheadandbloodybones, a traditional bogeyman. This terrifying, crouching figure, who is feared by children, appears on stage as Lily's baby is born and is intent on replacing the baby with a changeling (Meaker). Josie later alludes to this myth when she suspects that Lily's baby is a changeling: "That's not your baby. They've put one of theirs and taken yours off. . . . They'll keep yours down there. It makes them stronger. They'll breed from it" (Churchill, *The Skriker* 38). According to the traditional myth, "the children left behind, the changelings, were always malformed" (Phoenix n. pag.). The play's suggestion of the risk of congenital disability dramatises how the agency of the Skriker, figuring our toxic environment, hazardously affects other material bodies.[1]

In *The Skriker*, the new baby and, in particular, the baby's birth defects emphasise "not only . . . how various substances travel across and within the human body but how they *do* things – or often unwelcome or unexpected things" (Alaimo, *Bodily Natures* 146). In a context of high ecological risk and toxicity, according to Alaimo, some substances or xenobiotic chemicals "deviate," in a hazardous way, from their normal state and change into unknown and uncontrollable agents. She writes that "the human body can never be disentangled from the material world, a world composed of emergent, entangled biological creatures as well as a multitude of xenobiotic, humanly made substances" (24). The body, which is entangled with such ostensibly harmless chemicals, is "the chemically reactive body" that deviates from standard measures of scientific knowledge (138). These incalculable, invisible, and destructive forms of violence or "slow-motion toxicity" illustrate what Rob Nixon calls "slow violence" because they take the form of "scientifically convoluted cataclysms in which casualties are postponed, often for generations" and "dispersed across time and space" (2–3). Lily's dead child and deformed great-great-granddaughter embody this form of slow violence. Rather than subscribe to the narrow perspective that birth defects are often due to genetics, Alaimo suggests that it is not possible to distinguish between the destructive consequences of environmental toxins and the development of fetuses in the human body (104).[2] In the play, Lily's maternal body is a polluted "material self" or an example of the chemically reactive body that is inextricably and "intra-actively" interconnected with swarms of nonhuman agencies and poisonous substances in the deteriorated and deformed environment represented by the Skriker and fairies.[3] This trans-corporeality has made Lily's body the host of strange strangers. As I shall explain, Lily's travel to the underworld shows how the lives of future generations are unknowingly affected by such a poisoned material body.

In the final phase of the play, both Lily and Josie's bodies are constantly invaded by the Skriker, and the women cannot resist its penetration into their lives. In an attempt to save Lily and her baby and to pacify the Skriker and keep it "sated seated besotted," Josie ironically promises to do terrible things: "Just leave it to me. You don't have to do anything. . . . You won't do anything to Lily?" (Churchill, *The Skriker* 48). She starts to commit violent crimes, going "further and murder in the dark" (49). In a deeper sense, Lily decides to offer herself to the Skriker, who, in the guise of a young man, a young woman, and a very ill old woman, has begged for Lily's love and support. Lily, who tries to keep her baby and Josie safe while also helping the Skriker, trusts that her nightmarish journey to the underworld will be a transitory event that will allow her to bring everyone to salvation. She imagines she will return to the upper world and rejoin her child at the moment when she abandoned her: "if it's what Josie did I'll be back in no time" (51). However, during the time when Lily is left alone by the Skriker before going to the netherworld, the haunting presence of Black Annis onstage, though Lily cannot see her, compounds the dangers that await Lily's child. According to folklore, this blue-faced crone hides in an oak in order to ambush young children who wander too far from home before drinking their blood and eating their flesh (Meaker). In this way, the play implies that Black Annis ambushes Lily's child when she leaves her behind to go to the underworld.

Getting "tricked tracked wracked" (Churchill, *The Skriker* 52), Lily discovers that the baby has died and that she has lost a hundred years. Her efforts to save the world have been futile. The Skriker, who is no longer old but now full of energy as at the beginning of the play, tells Lily that she has moved not to the underworld but to the fatal future: "Lost and gone for everybody was dead years and tears ago, it was another cemetery, a black hole hundred yearns" (51–2). In a gloomy scene where a blackbird sings, Lily appears "like a ghastly" to an Old Woman, who is her granddaughter, and a Deformed Girl, who is the woman's granddaughter. The Skriker narrates the scene in a closing monologue:

> And this old dear me was Lily's granddaughter what a horror storybook ending. 'Oh I was tricked tracked wracked,' cried our heroine distress, 'I hoped to save the worldly, I hoped I'd make the fury better than she should be.' And what would be comfy of her now? She didn't know if she ate a mortal morsel she'd crumble to dust panic.
>
> (52)

The play's future generation of dead and malformed children might be seen as a ramification of earthly processes occurring in a toxic world. Lily's poisoned body has crumbled to "dust panic." Her deformed great-great-grandchild, the result of the contaminated food Lily ate, "*bellows wordless rage*" at her (52) when she understands that Lily is from the distant past and responsible for her miserable situation. The Old Woman, Lily's granddaughter, however, is sorry for Lily. She believes that people in the past did not understand what they did and that their stupidity explains their passivity: " 'Oh they couldn't helpless . . . they were stupid stupefied stewpotbellied not evil weevil devil take the hindmost of them anyway'" (52). In spite of this, the great-great-grandchild cannot forgive the irresponsibility of past generations and hates "the monstrous" – that is, Lily.

In *Risk Society: Toward a New Modernity* (1992), Ulrich Beck underscores how future generations are affected by present toxicity:

> [T]he ecological and high-tech risks that have upset the public for some years now . . . are no longer tied to their place of origin, the industrial plant. By their nature they endanger all forms of life on this planet. . . . They outlast generations. The affected even include those not yet alive at the time or in the place where the accident occurred but born years later and distances away.
>
> (22)

The long-term effects of human-induced ecological threats cannot be entirely controlled and understood. In this context of global risk, Lily's deformed descendant appears to be the victim of anthropocentric actions and behaviours. She offers a glimpse of the nightmarish reality of a distorted and poisoned environment in which pregnant women unwittingly come into trans-corporeal contact with toxins. This apocalyptic image of the future of the human world prefigures Churchill's 2006 libretto, *We Turned on the Light*, in which she lists human activities and their aftermath: "We wrapped the food in plastic/And saw the bears starving" and "We chopped down a forest/And heard a child choking" (Churchill, *We Turned on the Light* n. pag.). A modern consumerist lifestyle poisons food (the wrapping of food in plastic) and diminishes natural resources (the chopping down of a forest) while bringing about the extinction of nonhuman animals (the bears starving) and damaging human bodies of current and future generations (a child choking).

Being permeable and thus vulnerable to the sick environment and its harmful substances, Lily's body is now further poisoned by the human food that the Old Woman offers her at the very end of the play. Lily's toxic body is finally bound to toxic foods, and the Skriker announces, in a cheerful manner, that Lily has become a "dustbin":

> [']Have a what drink, wrap her in a blanket out, have a sandwhich one would you like?' So Lily bit off more than she could choose. And she was dustbin.
>
> *The* OLD WOMAN *holds out some food and* LILY *puts out her hand to take it.*
>
> (Churchill, *The Skriker* 52)

The toxic environment has become the very material of Lily's body. The scene dramatises a tragic moment in which her body is diminished by all of the toxins she has absorbed during her entire life. At this point, the Passerby, who has been dancing uninterruptedly throughout the play, stops dancing. This might well refer to the end of "human history" (Adiseshiah 296) or, in my view, the end of anthropocentrism and the start of the Anthropocene, in which humans have lost control of the changes for which they are responsible. The play's final scene provokes a new sense of tragedy that undermines the humanist bias of theatre. According to Diamond, this new "posthuman tragedy" is founded upon "the impersonal materiality of the natural world" rather than "traditional tragedy's grand anthropocentric designs" ("Tragic Materialism" 756). Lily's demise is marked by impersonal materiality and by the fact that the Passerby stops dancing, suggesting how, in Diamond's words, the play's decentered tragic framework "seeks the more intimate ground of shared materiality" (756). In this sense, I argue that the play models a non-anthropocentric and shapeshifting dramaturgical praxis for designating a disembodied human self and expressing a trans-corporeal world in which multiple networks of bodies and material agencies are entangled.

It is hardly surprising that ecological changes and biological transmutations displace human beings due to the manner in which humans have destroyed the planet, even if such displacement also affects the poor and weak, who arguably bear less responsibility for ecological disasters. The Skriker and the fairies take their horrifying revenge on the present and future world of Lily and Josie. Although the two women, one kind and the other unkind, represent different responses to the Skriker's offers, they intend to save the world from the spirit's vengeful power. However, the poor, innocent, kind Lily suffers from the unjust consequences of helping the Skriker. The Skriker explains to Lily that she has chosen her out of all other human beings because "You're the only good person I've ever met. Everyone else has tried to destroy me. But you wish me well. You wouldn't deny that" (Churchill, *The Skriker* 45). Nevertheless, Lily, who voluntarily accepts the Skriker's offer to go to the underworld for the good of the future, eventually finds herself in "another cemetery" where her future is lost. While the consumption practices of first world nations may be directly responsible for much of the devastation of ecological disasters, it is more frequently those in poorer nations who suffer the environmental fallout. According to Alaimo, "[a]s a particularly vivid example of trans-corporeal space, toxic bodies insist that environmentalism, human health, and social justice cannot be severed" (*Bodily Natures* 22). Although toxins and chemicals affect everyone, those most vulnerable to their effects are children, pregnant women, and developing fetuses (83). Those effects shape the pregnant Lily, Josie who killed her own child, and the children of the following generations. Thus, the play urgently reminds us of global environmental injustice: environmental degradation is not equally distributed among genders and classes. In this sense, the environment is not a force for humanity's liberation and betterment.

The onto-epistemological shift enacted by Churchill's dark picture of human-nonhuman interaction results in a sort of material ethics that is barely recognisable: it demands that "we inquire about all of the substances that surround us, those for which we may be somewhat responsible, those that may harm us, those that may harm others, and those that we suspect we do not know enough about" (Alaimo, *Bodily Natures* 18). In this swirling and ever-changing environment, the interaction among human and nonhuman forces is complex and far beyond what an exclusively anthropocentric approach could reveal. Considering this ecological context, Lily and Josie are not determined and bounded entities but rather material selves who are inseparable from the wider material flows of nonhuman actors. This situation ruptures any distinctive mode of self-knowledge, highlighting the permeability of the "material self" to a variety of scientific, "biological, economic, and industrial systems that can never be entirely mapped or understood" and fostering "a new genre of autobiography," or what Alaimo calls a "material memoir" (23). This picture of the trans-corporeal state of the material self is epitomised in the shapeshifting space of the play in which human health is inextricably entangled with environmental health issues, and human individual and societal instability coexists with a damaged environment. Such a profound shift in selfhood and self-knowledge dissolves the human protagonist and makes us think of a suitably humble response to the Anthropocene, which might be acknowledged as a different autobiography of humanity. Although humans are now labeled geological agents, the self of our material memoirs – in other words, our posthuman subjectivity – needs to be redefined in conjunction with the changes of this age. In that light, the play's trans-corporeal space dramatises a posthumanist notion of ecological awareness and environmental ethics.[4] A performative and non-anthropocentric ethical action resists any boundaries and disconnection, revealing that our responsibility is not to an outside Nature in the background but to a material world of trans-corporeality, of interrelated agencies, that resides inside our bodies and of which we are a part.

2 A materialist design for political agency in *Far Away*

Premiering at the Royal Court Theatre Upstairs in London in November 2000 and directed by Stephen Daldry, *Far Away* dramatises the materiality of human-nonhuman relationships throughout its three acts in a way that anticipates a post-anthropocentric future. The first act introduces a frightened young girl, Joan, who asks her aunt, Harper, probing questions about disturbing sights and sounds in the farmhouse shed. In the second act, several years have passed, and Joan works as a designer in a hat factory, creating exotic hats to be worn by parades of chained prisoners who are sentenced to execution. The first two parts of the play serve as precursors of the final sequence: the dark and mysterious environment of Act One and the images of human suffering in Act Two prefigure a mad fantasy image of all-out ecological warfare among humans, animals, plants, the weather, the landscape, and geology. The hazardous and grotesque landscape of ecological dystopia that emerges in Act Three blurs the distinction between lifeforms and other material entities. The landscape embodies an ontological narrative of the vitality of all sorts of bodies and entities interacting with each other in unpredictable ways; at the same time, each has its own particular story and meaning.

A space of vibrant matter: things as actants in dramatic assemblages

After examining how the hats in Act Two materialise an unbalanced but mutual connection among bodies in the context of a corrupt system, this section focuses on the vital materiality of Act Three. Such an exploration raises the question of how the audience's democratic, socio-political, economic, and ecological analysis of this scene of violence might be shaped by the way in which the play gives dramatic form to the agency of things – which are not recognisable as alive but have effects. In this case, how does the audience respond to the role of the hats, which are represented here in grotesque and overblown proportions?

In the second act, taking place in a hat makers' factory, Joan and a young man named Todd make hats intended for prisoners' executions: "A procession of ragged, beaten, chained prisoners, each wearing a hat, on their way to execution" (Churchill, *Far Away* 30). Day after day – and scene after scene – the hats grow larger and brighter, until they are described as "enormous and preposterous" (30). Each week, we learn, one hat is selected as the most decorative one in a hat competition and is kept and displayed in a museum; the rest of the hats are burned along with the corpses of the prisoners. Mass murdering is reduced to a show trial, with which the hat-making industry is sustained (see fig. 2.5). A note in the foreword to the play specifies this progress in the magnitude of horror and violence: "The Parade: five is too few and twenty better than ten. A hundred?" (8). Todd and Joan, who negate the outward reality of the death parades and are indifferent to this deeply brutal and corrupt environment mandated by a capitalist system, are concerned with the creation of the hats and their aesthetic and artistic merits:

JOAN It seems so sad to burn them with the bodies.
TODD No, I think that's the joy of it. The hats are ephemeral. It's like a metaphor for something or other.
JOAN Well, life. (31)

As the ragged prisoners march towards death, the outlandish proportions and preposterous shapes and colours of the hats are highlighted at the expense of the tortured humans wearing them, who are used as means for showing such absurd magnificence. Rather than merely functioning like a metaphor for life, the hats are intimately connected with a dystopian reality and its brutal killing, heightening the issue of agency. According to Marissia Fragkou in *Ecologies of Precarity in Twenty-First Century Theatre* (2018), "dispossessed human life made an appearance through [an] object" (86). It might seem that the hats are dead, inanimate, and passive objects. However, drawing on Jane Bennett's notions of "thing-power" or "vibrant materialism," which remind us of "the force of things" (1) and "the curious ability of inanimate things to animate, to act, to produce effects dramatic and subtle," I characterise the hats as "vivid entities" (5–6). However, such a material encounter occurs without centralising the agency of the hats or trivialising the oppression of the humans who perish alongside the hats, but through interaction among the hats and silenced human bodies – that are deprived of identity – in the material world. In her article "Different Hats" (2003), Chaudhuri sees the representation of objects through the medium of theatre as a recognition of "our 'co-life' with things" in a way that engenders an ecological awareness of the interconnectedness of humans and nonhumans (134). In a broader sense, by making a difference and producing effects within the network of relationships of which they are a part, the hats "become more than mere objects" (Bennett, *Vibrant Matter* 13).[5] They materialise humanity's disregard for humanity. In this way, they bring into focus, in Bennett's words, "the shared materiality of all things," minimising the difference between subjects and objects (13).

The brutal, threatening, and disordered second act can be discerned in the concretised "agentic capacity" of the hat, to use Bennett's term, with its creative and destructive sides. As Bennett asserts, "the pure power of *a* life" or a matter "can manifest as beatitude, or as an unspeakable, sheer violence" (53). The glamorised hats, as aesthetic forms, become complicit with a system

Figure 2.5 Far Away. The Cleveland Play House in Ohio, PavelDobrusky Multimedia LLC, 2004. Directed by Peter Hackett. Photo by Roger Mastroianni.

of barbarism and repression, producing real effects of horror. According to Chaudhuri, through *Far Away*'s hats, the playwright shows "a 'dire threat' to life of things," and the hats are imbued with "a deathliness of their own" ("Different Hats" 134). By focusing on the hats, readers and the audience move away from a narrow vision in which characters and their features are seen as distinct categories, and towards seeing how material things shape characters and connect them to the system of brutality that the characters inherit. Given this dynamic and interactive environment, we become aware that the ecology of everyday life cannot be limited to our anthropocentric activities and perspectives. Rather, human actions take place in a complex mesh of materials in which everything affects everyone else. By representing the possibility of a material relationship between things and humans, the play's hat scene takes a step towards the wider ecological insight of Act Three, in which an indivisible continuum of manifold bodies is on the move.

Act Three famously depicts a shocking future world in which humans and nonhumans are represented as having an unusual and unexpected kind of agency. In this darkly imagined world, various animal species have begun entering into alliances with human nations, creating a global landscape of pan-species warfare. In the 2010 production directed by Simon Godwin at the Theatre Royal at the Bristol Old Vic, the sound of eerie birdsong in Act One and the abrasive sounds of tearing and sewing in the factory in Act Two are replaced by the death rattles of animals and the menacing caws of crows (Gardner). Such dark reverberations around the stage intensify the dreadful and mysterious atmosphere of the play. From the very first strange images of aggressive wasps and hostile butterflies, it is evident that the terror and aggression among humans in the first two acts have now been transferred into the natural environment. Harper and Todd provide an extensive list of uncertainties caused by not knowing which side different beings are on. For example, Harper says: "The cats have come in on the side of the French" (Churchill, *Far Away* 35), and elsewhere she says: "Mallards are not a good waterbird. They commit rape, and they're on the side of the elephants and the Koreans" (39), whereas "the elephants went over to the Dutch" (40). Joan also narrates her vision of the violence, which has intensified from the domestic corruption she witnessed as a child to mass-murdering system of corrupt hat industry in Act Two into global degradation. She seeks refuge in the middle of a disastrous inferno in which everything has taken a side and been "recruited" in response to some unknown instability that may or may not be a futuristic ecological crisis. Nothing and no one can be trusted (see fig. 2.6). Peter Hackett's direction and Pavel Dobrusky's scenic design for *Far Away*'s production at the Cleveland Play House in April 2004 endowed the play with creative physicality. The play's "storybook setting" was made out of a colourfully painted backdrop composed of wooden slats that sweep upward (Heller n. pag.). Its first scene of a painted show-drop, which depicts an idyllic countryside, soon appears to be illusory as it shifts to a dark vision of what Diamond calls a "posthumanist" chaos.

Figure 2.6 *Far Away*. Joan in the middle of a disastrous inferno. The Cleveland Play House in Ohio, Pavel Dobrusky Multimedia LLC, 2004. Directed by Peter Hackett. Photo by Roger Mastroianni.

In this description of the natural world, nonhumans have agency as humans do, but in an ugly and terrifying way. These surreal images of animals – cats killing babies and butterfly wings violently smothering humans – present a desentimentalised view of the agency of animals and result in a new kind of "ecological thought." The ecological thought, as Morton argues, is not just about affirmative images of the environment but includes "ugliness and horror" (*Ecological Thought* 17). In his view, "dark ecology" compels "our compassionate coexistence [with nonhuman animals] to go beyond condescending pity" (17). Such ecological thinking suggests a democratic ecology in which each entity has its own agency in a non-hierarchical environment. Darkness and violence in ecological thinking can be considered a contribution to this ecological democracy because the more we are aware of risk and threat, the more we acknowledge that we are enmeshed in and affected by the ecological changes happening on a global scale. The play's dark space of vitality is founded upon a reciprocal identification that may decentre anthropocentric dominance and control and change how we think about our relationships with nonhumans. Further, it would seem that Act Three actively plays on the mode of eco-horror in order to wake us up to a darker ecological reality, in which human beings are instigators of chaotic changes in the environment but are certainly not in control of those changes.

Joan narrates a dynamic network of both intimately and hostilely interconnected entities in a more extreme way while taking shelter back at her aunt's home. Joan indifferently and unsympathetically kills two cats and a child under five on her way to Harper's house, indicating that she has normalised every kind of brutality. Joan and Todd are ethically desensitised and unsympathetic to their environment. The disruption of the conventional distinctions between humans and nonhumans can be a means by which to heighten audiences' and critics' awareness of ecological interactions while it can also be a means of criticising the mainstream belief in an ethical division between humans and nonhumans. In this regard, Sheila Rabillard argues that if we judge the characters to be morally desensitised because they do not differentiate killing humans from killing nonhumans, "the presumed distinction between the status of human and animal . . . may itself be a means by which the audience makes its own familiar acts of violence against the natural world acceptable" (101). In addition to the brutality of the characters' actions and their dark description of the natural environment, the play's juxtaposition of the killing of animals with the killing of humans implies a horizontal rather than hierarchical relation between humans and nonhumans. In this ecological community, things, humans, and nonhumans are constantly shifting. This relation does not mean, of course, that humans and nonhumans are equally entitled (Bennett, *Vibrant Matter* 104), but such a perspective has an ethical outcome: it makes humans more receptive to nonhumans as active participants in a public and shared ecological space.

Joan continues her description of the environment as it embodies the vital materiality of a wide range of nonhuman forces acting alongside humans. In a terrifying monologue, she tells us about her fear of everything in her surrounding environment – not only the birds and rats but also the weather and thunderstorms, along with inanimate things:

> It wasn't so much the birds I was frightened of, it was the weather, the weather here's on the side of the Japanese. There were thunderstorms all through the mountains, I went through towns I hadn't been before. The rats are bleeding out of their mouths and ears, which is good, and so were the girls by the side of the road. It was tiring there because everything's been recruited, there were piles of bodies and if you stopped to find out there was one killed by coffee or one killed by pins, they were killed by heroin, petrol, chainsaws, hairspray, bleach, foxgloves, the smell of smoke was where we were burning the grass that wouldn't serve.
>
> (Churchill, *Far Away* 43)

War is now total: readers and the audience encounter an image of the entire planet mobilised against itself. The birds are to be feared, while the weather is dreadful and can no longer be trusted. Meanwhile, inanimate things are animated by the capacity to kill, while "we" also take a side, asserting our hostility to the grass. The terror that Joan speaks of is not limited only to particular species, humans, and regions – as she says, "everything's been recruited" (43).

This passage's language of violence, which Chaudhuri considers essential to the play's representation of an ecological mess, "should not obscure another, more potentially creative feature of the world depicted here: the crossing of species and even material boundaries, resulting in novel alliances . . . and an implicitly collaborative account of earthly existence" ("Anthropo-Scenes" 315). In sync with Chaudhuri, I argue that this passage's language reveals another experimental aspect of the material world narrated in Act Three, which foregrounds not merely a space of "interspecies war" (Arons and May, "Ecodramaturgy" 194) but also a collective and communal landscape which articulates the entanglement of humans, nonhuman animals, and inanimate matter in a messy and complicated world of materiality. The current world is experienced through fears about such entanglement and its unpredictable materiality. In this way, the play dramatises the idea of "ecophobia," to use Simon Estok's term, demonstrating fears of the loss of predictability in such an increasingly complex ecological mess.

Such a surreal but not distant narration of the physical world brings this question to mind: What difference would it make to our ecological awareness if nonhuman things were considered, in Bennett's words, not simply passive and stable "resource[s]" or indices of "human mood, action, meaning, agenda, or ideology," but rather "potentially forceful agents" that participate in the play's global war (*Vibrant Matter* x, viii)? To answer this question, I consider how the play resonates with Bennett's description of the material objects she encounters while taking a walk. She is struck by:

> the materiality of glove, the rat, the pollen, the bottle cap, and the stick started to shimmer and spark . . . because of the contingent tableau that they formed with each other, with the street, with the weather that morning, with me. For had the sun not glinted on the black glove, I might not have seen the rat; had the rat not been there, I might not have noted the bottle cap, and so on. But they *were* all there just as they were, and so I caught a glimpse of an energetic vitality inside each of these things, things that I generally conceived as inert.
>
> (4–5)

Bennett's insistence on matter's vitality and its engagement in an assemblage is strikingly delineated in the theatrical space of *Far Away*. Each agent in the eco-global warfare of Act Three has a vital force, is continuously pulsing with life, and insists on its participation in the ongoing exchanges that encompass the biosphere. Of course, aliveness does not always produce a positive and desired effect on other bodies. For example, a threatening dynamism or violent vitality is awakened when agents become destructive as they are affected by the invisible toxic materials in the environment or by other ecological changes. The dynamic materiality of the play is thus composed of manifold human and nonhuman actors, or "actants" to use Bruno Latour's term, that participate in global ecological terror and unfold new possibilities and new actantial functions.[6] Bennett invokes Latour's concept of the actant to oppose the subject-object binary: an actant is "a source of action. . .; it is that which has efficacy, can do things, has sufficient coherence to make a difference, produce effects, alter the course of events" (Bennett, *Vibrant Matter* viii). In sync with this idea, I argue that, in the play, the animate and inanimate things that Joan lists are different types of actant that are figured as having thing-power, are part of a network of materially heterogeneous actants, and are capable of affecting and being affected beyond the preview of anthropocentric action. In this light, human and nonhuman actors in a theatrical work might be considered as actants insofar as their agency and efficacy is manifested through the interaction of many characters, bodies, forces, and dramaturgical and material elements.

The materiality of each actant in the play is also defined through its inextricable connection with manifold other actants and in response to many social, political, and ecological factors. Thus it is not possible to discern the causality or origin of an event or action because they are inherently affected by each other. Bennett's concept of "distributive agency" is useful here because its focus is not to "posit a subject as the root cause of an effect" but to identify "a swarm of vitalities at play . . . and the kind of relations that obtain between its bits" (31). Since agency is distributive, there is no linear or "efficient" causality but rather an "emergent" or "fractal" one: "If efficient causality seeks to rank the actants involved, treating some as external causes and others as dependent effects, emergent causality places the focus on the process as itself an actant, as itself in possession of degrees of agentic capacity" (33). This perspective illuminates the play: rather than ranking human and nonhuman actants and linking them to external causes and dependent effects, the uncertainty of the play's environment suggests an emergent causality in which a "swarm" of actants pass through, intervene, and "intra-act." No single agent is in charge of this tumultuous situation, and no one can find and foretell the cause or the outcome of this unexpected coming together of various bodies. The play's performance of emergent causality invites us to consider the agency of an ecological "assemblage," which includes not only "the vitality of a discrete thing" but also "vitality as a . . . function of the tendency of matter to conglomerate or form heterogeneous groupings" (Bennett, *Vibrant Matter* xvii).[7] Bennett's notion of assemblage, borrowed from Gilles Deleuze and Félix Guattari, identifies the dissemination of matter across a broad and heterogeneous range of nonhuman forces or actants, such as "trash, bacteria, stem cells, food, metal, technologies, [and] weather" rather than "a capacity localized in a human body or in a collective produced (only) by human efforts" (Bennett, *Vibrant Matter* 23). In this way, the agency of matter is always dependent on a sense of relationality instead of being limited to any centralised source of authority.

The play's final act imagines a performative context in which the agentic assemblage of a heterogeneous series of human and nonhuman actants creates emergent effects and unpredictable transformations outside of human control. Of course, while the ecological mesh might democratise agency, it does not eliminate humanity's badly informed activity. The main challenge of the Anthropocene is that humans can affect, not control. Given the fact that human activities reverberate across a variety of beings in the ecosystem, the play demonstrates nonhuman actants that have agencies that cannot be entirely dominated or assimilated into human categories or plans. Hence, the play's narration of interaction among various actants tells us about the rise of strange and dark processes of uncertainty as well as a new conception of humanity. For example, Joan's description of nightmarish realities outlines this ongoing uncertainty: she asserts that the "Bolivians are working with gravity" while "we're getting further with noise and there's thousands dead of light in Madagascar," and then asks about "who" is "going to mobilise darkness and silence?" (Churchill, *Far Away* 43–4). Through such pronouncements, Churchill alludes to the ways in which humans have become a potent environmental force by altering the climate and landscapes as well as flows of species, energy, and materials. However, our immersion in natural systems extends scales of perception, demonstrating our own entanglement in material dynamics we cannot command. For example, the Bolivians control and manipulate gravity, creating out of it the next weapon of mass destruction. Meanwhile, others use light to kill thousands in Madagascar. Forms of violence against the planet are thus condensed into the fundamental question, "Who's going to mobilise darkness and silence?" (44). This question might well evoke the current era of the Anthropocene and our new consciousness of humankind's agency. Our activities have mobilised the earth system as a whole so that all actants with thing-power are now in play in an uncertain way. The play contributes to the drama of the Anthropocene, which challenges traditional methods and established protocols of knowledge production in the humanities, by provoking an awareness that the human has become a material actant in a swarm of vital materialities.

Given such an unmanageable complexity, it is not possible to say who or what is at the centre of the play's ecological assemblage: People from different nations? The animals? The weather? Hats and the other things that Joan lists, such as coffee, pins, heroin, petrol, chainsaws, hairspray, bleach, foxgloves, light, and sound? The play stages the irreducible roles of things. Each of these materials is a small agency with a distinct effect and capacity, though they are not arranged in a hierarchical order and their agency is distributed across a heterogeneous collection of matter. Hence, Churchill's play assembles a wide-ranging collection of humans, nonhuman animals, and things into a narrative sequence that hinges on "a vital materialist theory of democracy" which "seeks to transform the divide between speaking subjects and mute objects into a set of differential tendencies and variable capacities" (Bennett, *Vibrant Matter* 108). In this incalculable environment, the Moroccans, the Canadians, the Venezuelans, engineers, chefs, children under five, musicians, Russian swimmers, Thai butchers, dentists in Latvia, and Spanish computer programmers are united ontologically with butterflies, wasps, cats, crocodiles, deer, weather, light, water, coffee, foxgloves, petrol, and so forth – though these performative warring alliances are against others and against each other. While the play's violence and global war dramatise the agency of things and nonhumans, I contend that it also evokes democracy in that the play gives voice to passive objects and nonhuman agents that act performatively, intensely, and heterogeneously, though unequally, in creating a material world descending into terror.

A moment of uncertainty that results from this obscure swarm of activities occurs at the end of the play, when Joan describes her fear crossing a river:

> But I didn't know whose side the river was on, it might help me swim or it might drown me. In the middle the current was running much faster, the water was brown, I didn't know if that meant anything. I stood on the bank a long time. But I knew it was my only way of getting here so at last I put one foot in the river. It was very cold but so far that was all. When you've just stepped in you can't tell what's going to happen. The water laps round your ankles in any case.
>
> (Churchill, *Far Away* 44)

Forced to cross the river, Joan steps into the cold, running water. Her last sentence imagines a world where "[t]he water laps round your ankles in any case," indicating that survival is contingent on the incalculable and horrible entanglements of human and nonhuman actants and forces beyond our perception. The immersive and weird agentic capacity of water illuminates ecophobic performance because the fear of water and its unpredictability becomes the fear of its nonhuman agency (Estok). This knowledge of the collective nature of human life, which is itself a diverse assemblage, and its intimacy with nonhuman forces, which might seem to be totally unrelated or even hostile to it, brings about a radical shift in the anthropocentric conception of humanity. It reveals how we are deeply immersed in unprecedented eco-global assemblages where "you can't tell what's going to happen," as Joan says. In the current world of global disequilibrium, a world out of control, subjectivity is disrupted. Although humans and nonhumans can each perform in their own independent ways, they are deeply affected by each other. The play's dark ecology comes out of such uncertainty.

This chapter has offered insights into Anthropocenic theatre by foregrounding Churchill's ways of addressing "new forms of narrative agency" (Iovino 139) and setting out a theatrical space constituted by the distributive agency and trans-corporeal interactions among human and nonhuman actants. The plays' embodiment of the human characters as material entities that are entangled with a vast, mysterious, and vital assemblage of agential forces leads us to reconsider our fundamental ethical and political accountability. Although Churchill's landscapes are dark and full of strangeness and mystery, this uncertainty gives a new insight into the ecological crisis of our time. It also suggests a different way of performing the human story by listening to the stories of the nonhuman environment.

Notes

1 In the article "'A Licence to Kill': Caryl Churchill's Socialist-Feminist 'Ideas of Nature'," Elaine Aston explores the play from an ecofeminist perspective, focusing on the interchanges between maternal and ecological worlds. For Aston, "Churchill's vision is of a world in which the eco-microsystem of mother and foetus is damaged by the eco-macrosystem of mother and planet" (Giannachi and Stewart 43).
2 Stacy Alaimo refers to Sandra Steingraber's book *Having Faith: An Ecologist's Journey to Motherhood* (2003), in which Steingraber "dramatizes trans-corporeal processes, as various toxins flow through the environment into water, air, and food, through the mother's body, and finally, into the developing fetus and the nursing infant" (*Bodily Natures* 104). In this book, Steingraber describes her body as an environment and criticises the lack of consideration given to the impact of environmental threats to pregnancy and fetuses. As she argues, "protecting the ecosystem inside my body required protecting the one outside" (qtd. in Alaimo 104). Steingraber also warns about "teratogenic effects" of many common household substances, as the causes of birth defects, and how "her body, the habitat for her developing child, is inextricably connected to the wider world" (104).
3 According to Karen Barad, "*Matter is substance in its intra-active becoming – not a thing but a doing, a congealing of agency*" (*Meeting the Universe* 151). Intra-activity is different from inter-activity due to its rejection of the existence of "things" prior to their relations.
4 "Even though trans-corporeality emerged from environmental health and environmental justice movements, which focus on (post)humans, the concept extends toward all species who find themselves at the crossroads of body and place. The posthuman ethics of trans-corporeality insists that even the most routine human activities, such as purchasing plastics, impact human and nonhuman lives across vast geographic and temporal scales, extending even to the bottom of the sea" (Alaimo, "Trans-corporeality" 437).
5 "[I]t is not that individuated objects are willful subjects; but they can be powerful actants in operation with others. The locus of agency is in practice always a human-non-human collective" (Bennett, "Vibrant Matter" 448).
6 In *The Pasteurization of France*, Bruno Latour says an actant can be "a storm, a rat, a rock, a lake, a lion, a child, a worker, a gene, a slave, the unconscious, or a virus" (192).
7 The idea of ecological assemblage also reflects Bruno Latour's "Actor-Network Theory," which proposes the constant interchanges between vibrant actants of various categories.

References

Adiseshiah, Siân. "The Dramatisation of Futureless Worlds: Caryl Churchill's Ecological Dystopias." *Dystopia(n) Matters*. Ed. Fátima Vieira. Newcastle: Cambridge Scholars, 2013.294–305. Print.

Alaimo, Stacy. *Bodily Natures: Science, Environment, and the Material Self*. Bloomington: Indiana UP, 2010. Print.

———. "Sustainable This, Sustainable That: New Materialisms, Posthumanism, and Unknown Futures." *PMLA: Publications of the Modern Language Association of America* 127.3 (2012): 558–64. Print.

———. "Trans-corporeal Feminisms and the Ethical Space of Nature." *Material Feminisms*. Eds. Stacy Alaimo and Susan J. Hekman. Bloomington: Indiana UP, 2008.237–64. Print.

———. "Trans-corporeality." *Posthuman Glossary*. Eds. Rosi Braidotti and Maria Hlavajova. London: Bloomsbury, 2018.435–8. Print.

Amich, Candice. "Bringing the Global Home: The Commitment of Caryl Churchill's *The Skriker*." *Modern Drama* 50.3 (Fall 2007): 394–413. Print.

Arons, Wendy and Theresa J. May. "Ecodramaturgy in/and Contemporary Women's Playwriting." *Contemporary Women Playwrights: Into the Twenty-First Century*. Eds. Penny Farfan and Lesley Ferris. New York: Palgrave Macmillan, 2013.181–96. Print.

Aston, Elaine. " 'A Licence to Kill': Caryl Churchill's Socialist-Feminist 'Ideas of Nature'." *Performing Nature: Explorations in Ecology and the Arts*. Eds. Gabriella Giannachi and Nigel Stewart. Oxford: Peter Lang, 2005.165–78. Print.

Bane, Theresa. *Encyclopaedia of Fairies in World Folklore and Mythology*. Jefferson: McFarland, 2013. Print.

Barad, Karen. *Meeting the Universe Half Way: Quantum Physics and the Entanglement of Matter and Meaning*. Durham, NC: Duke UP, 2007. Print.

Beck, Ulrich. *Risk Society: Towards a New Modernity*. Translated by Mark Ritter, London: SAGE Publications, 1992. Print.

Bennett, Charlie. "Review: The Skriker at Royal Exchange Theatre, Manchester International Festival." *Aesthetica Magazine*. Aesthetica, 9 July 2015. Web. 02 Mar. 2016. <www.aestheticamagazine.com/review-skriker-royal-exchange-theatre-manchester-international-festival/>.

Bennett, Jane. *Vibrant Matter: A Political Ecology of Things*. Dunham and London: Duke UP, 2010. Print.

———. "Vibrant Matter." *Posthuman Glossary*. Eds. Rosi Braidotti and Maria Hlavajova. London: Bloomsbury, 2018.447–8. Print.

Briggs, Katharine Mary. *The Fairies in Tradition and Literature*. London and New York: Routledge, 1967. Print.

Campbell, John. *Margaret Thatcher Volume Two: The Iron Lady*. London: Random House, 2011. Print.

Chambers, Amy C. "The Skriker: Global Warming, Eco-Fairytales, and Science on the Stage." *The Guardian*. Guardian News and Media, 24 July 2015. Web. 02 Mar. 2016. <www.theguardian.com/science/the-h-word/2015/jul/24/skriker-global- warming-eco-fairytales-science-stage>.

Chaudhuri, Una. "Animals Rites: Performing beyond the Human." *Critical Theory and Performance*. Ed. Janelle G. Reinelt and Joseph R. Roach. Ann Arbor: The University of Michigan P, 2007.506–20. Print.

———. "Anthropo-Scenes: Staging Climate Chaos in the Drama of Bad Ideas." *Twenty-First Century Drama: What Happens Now*. Eds. Siân Adiseshiah and Louise LePage. London: Palgrave Macmillan, 2016.303–21. Print.

———. "Different Hats." *Theater* 33.3 (Fall 2003): 132–4. Print.

Chaudhuri, Una and Shonni Enelow. *Research Theatre, Climate Change, and the Ecocide Project*. New York: Palgrave Macmillan, 2014. Print.

Churchill, Caryl. *Churchill: Shorts: Short Plays*. London: Nick Hern Books, 1990. Print.

———. *Escaped Alone*. London: Nick Hern Books, 2016. Print.

———. *Far Away*. London: Royal Court, 2000. Print.

———. *Lives of the Great Poisoners*. London: Methuen Darama, 1993. Print.

———. *The Skriker*. London: Cox & Wyman Ltd, 1994. Print.

———. *We Turned on the Light*. Dir. Orlando Gough. BBC Proms. Prom., 29 July 2006. Web. 02 Nov. 2012. <www.bbc.co.uk/events/rgwrbp/series>.

Cousin, Geraldine. "Context of Earlier Plays by Caryl Churchill." *Essays on Caryl Churchill: Contemporary Representations*. Ed. Sheila Rabillard. Winnipeg: Blizzard Publishing P, 1998.189–205. Print.

Diamond, Elin. "Caryl Churchill: Feeling Global." *A Companion to Modern British and Irish Drama 1880–2005*. Ed. Mary Luckhurst. Malden: Blackwell, 2006.476–87. Print.

———. "Churchill's Tragic Materialism; or, Imagining a Posthuman Tragedy." *PMLA: Publications of the Modern Language Association of America* 129.4 (2014): 751–60. Print.

———. *Unmaking Mimesis: Essays on Feminism and Theatre*. London: Routledge, 1997. Print.

Drake, Olivia. "Hand-Made Puppets, Costumes in Theater Department's Skriker." *WesleyanUniversity*. Wesleyan University, 18 Nov. 2009. Web. 11 May 2016. <http://newsletter.blogs.wesleyan.edu/2009/11/18/wesleyan-theater-department- presents-the-skriker-nov-20-22/>.

Estok, Simon C. *The Ecophobia Hypothesis*. New York: Routledge, 2018. Print.

Fragkou, Marissia. *Ecologies of Precarity in Twenty-First Century Theatre: Politics, Affect, Responsibility*. London: Bloomsbury Methuen Drama, 2018. Print.

Gardner, Lyn. "Far Away." *The Guardian*. Guardian News and Media, 29 May 2010. Web. 08 Feb. 2014. <www.theguardian.com/stage/2010/may/28/far-away-review>.

Giannachi, Gabriella and Nigel Stewart. "Introduction." *Performing Nature: Explorations in Ecology and the Arts*. Gabriella Giannachi and Nigel Stewart. Eds. Oxford: Peter Lang, 2005.19–62. Print.

Heller, Fran. "'Far Away' is Theatrical Coup." *Cleveland Jewish News*. 8 Apr. 2004. Web. 02 Nov. 2016. <www.clevelandjewishnews.com/archives/far-away-is-theatrical- coup/article_8eb555f6-d01e-502d-99d9-9bd5b44ee439.html>.

Hudson, Julie. "'If You Want to Be Green Hold Your Breath': Climate Change in British Theatre." *New Theatre Quarterly* 28.3 (2012): 260–71. Print.

Iovino, Serenella. "Steps to a Material Ecocriticism. The Recent Literature About the 'New Materialisms' and Its Implications for Ecocritical Theory." *Ecozon@* 3.1 (2012): 134–45. Web. 25 June 2014.

Iovino, Serenella and Serpil Oppermann. "Introduction: Stories Come to Matter." *Material Ecocriticism*. Eds. Serenella Iovino and Serpil Oppermann. Bloomington: Indiana UP, 2014. Print.

Kritzer, Amelia Howe. "Damaged Myth in Caryl Churchill's The Skriker." *Dramatic Revisions of Myths, Fairy Tales and Legends: Essays on Recent Plays*. Ed. Verna A. Foster. Jefferson: McFarland, 2012.112–22. Print.

———. "Systemic Poisons in Churchill's Recent Plays." *Essays on Caryl Churchill: Contemporary Representations*. Ed. Sheila Rabillard. Winnipeg: Blizzard Publishing P, 1998.159–73. Print.

Latour, Bruno. "On Actor-Network Theory. A Few Clarifications Plus More Than a Few Complications." *Soziale Welt* 47.4 (1996): 369–81. Print.

———. *The Pasteurization of France*. Translated by Alan Sheridan and John Law, Cambridge, MA: Harvard UP, 1993. Print.

Meaker, Courtney. *The Skriker Dramaturgical Notebook*. Seattle: Ghost Light Theatricals, 2011. Print.

Midgley, Patrick. "Food and *The Skriker*: Consumption and Corruption in Caryl Churchill's Posthuman Fairy Underworld." *Conversations with Food*. Eds. Dorothy Chansky and Sarah W. Tracy. Wilmington: Vernon P, 2021.99–110. Print.

Morton, Timothy. "Dark Ecology: Philosophy in the Anthropocene." Humanities Without Boundaries. UW-Madison Center for the Humanities, Madison. 19 Oct 2012. Lecture.

———. *Ecology without Nature: Rethinking Environmental Aesthetics*. Cambridge: Harvard UP, 2007. Print.

———. *The Ecological Thought*. Cambridge: Harvard UP, 2010. Print.

Nixon, Rob. *Slow Violence and the Environmentalism of the Poor*. London: Harvard UP, 2011. Print.

Phoenix, Liza. "Changeling." *Encyclopaedia Chimerica: The Phoenixian Book of Creatures.* 2010. Web. 04 Dec. 2013. <www.lizaphoenix.com/encyclopedia/changeling.shtml>.

Rabillard, Sheila. "On Caryl Churchill's Ecological Drama: Right to Poison." *The Cambridge Companion to Caryl Churchill.* Ed. Elaine Aston and Elin Diamond. Cambridge: Cambridge UP, 2009.88–104. Print.

Rose, Carol. *Spirits, Fairies, Gnomes and Goblins: An Encyclopaedia of the Little People.* Santa Barbara, Denver and Oxford: ABC-CLIO, 1996. Print.

Tompkins, Joanne. *Unsettling Space: Contestations in Contemporary Australian Theatre.* London: Palgrave Macmillan, 2006. Print.

Wilson, Ann. "Failure and the Limits of Representation in *The Skriker.*" *Essays on Caryl Churchill: Contemporary Representations.* Ed. Sheila Rabillard. Winnipeg: Blizzard, 1998.174–88. Print.

Wolf, Matt. "A Damaged World in which Nature is a Weirdo Killer." *The New York Times.* The New York Times Company, 5 May 1996. Web. 07 Jan. 2014. <www.nytimes.com/1996/05/05/theater/theater-a-damaged-world-in-which- nature-is-a-weirdo-killer.html>.

3 Setting the stage for the end of the world

Stephen Sewell's *It Just Stopped* and Andrew Bovell's *When the Rain Stops Falling*

This chapter analyses two Australian plays, *It Just Stopped* (2007) by Stephen Sewell and *When the Rain Stops Falling* (2009) by Andrew Bovell, as examples of recent Australian theatre that deals with themes of ecological change and climate risk. In Australia, environmental humanities has its roots in a project called the "ecological humanities" at the Australian National University (ANU) in the late 1990s (Rigby, "Weaving the Environmental Humanities"). This project, which was closely associated with environmental history, environmental philosophy, and Indigenous studies, was led by scholars from a diverse range of disciplines in the humanities and social sciences, especially Deborah Bird Rose, Libby Robin, Val Plumwood, and Tom Griffiths. The study of ecocriticism in the Australian humanities has developed slowly and environmental issues were under-researched in the field of literary studies. As Kate Rigby notes, the first environmentally inflected monograph in Australian literary studies was Bruce Bennett's *An Australian Compass* (1991), followed by the 2001 issue of the Australasian Universities Language and Literature Association journal, *AUMLA*, dedicated to "Nature and the Environment."[1] Further, the foundation of the Association for the Study of Literature, Environment and Culture, Australia and New Zealand (ASLEC-ANZ) and the related journal *Australasian Journal of Ecocriticism and Cultural Ecology* (AJE), along with the Ecological Humanities section in the *Australian Humanities Review* (AHR) from 2004 to 2012, have allowed Australian ecocriticism to develop local and global perspectives.

In particular, a number of scholars have introduced new ecocritical accounts of Australian theatre and performance. For example, Peta Tait has published on the practice and theory of theatre and body-based art forms and performance, including animal performers and theatrical languages of emotion. In *Wild and Dangerous Performances: Animals, Emotions, Circus* (2011) and *Fighting Nature: Travelling Menageries, Animal Acts and War Shows* (2015), she presents performing practices from the point of view of the books' nonhuman actors, rather than from the typical human perspective. In her article " 'Beauty Tigress Queen': Staging the Thylacine in a Theatre of Species" (2015), Denise Varney underlines the term "theatre of species" in the Australian context. She considers the ethical and political thinking prompted by encounters with nonhumans in the contemporary performance *They Saw a Thylacine* (2013), by

DOI: 10.4324/9781003048749-3

writer-performers Justine Campbell and Sarah Hamilton, in order to reveal "how Australian drama and performance indexes Australian thinking about ecology, culture and nature" (11). These examples illustrate that ecocriticism today does not limit its focus to European and North American writing, and so Australian theatre and performance can be considered an important ecological context for this study.

In fact, what makes Australian theatre responsive to ecocritical reading are its representations of a settler colonial landscape in which the natural environment has become vulnerable to environmental degradation.[2] According to David Lindenmayer in *On Borrowed Time: Australia's Environmental Crisis* (2007), although Australia has a remarkable and singular biological environment, its environmental problems and massive losses of biodiversity are unique and extraordinary as well (35). As Lindenmayer writes, poor natural resource management strategies and colonial land-clearing and its aftermaths – land degradation, salinity, sedimentation, rising water levels, extinction of waterbirds, plants and animals – together with industry, urban expansion, the introduction of invasive species and altered fire regimes are imported practices that have caused the destruction of biodiversity and the environment over the past 220 years (40–6).

Further, Australia's contribution to the theatre of climate change marks Australian perspectives on global Anglophone drama in the framework of contemporary concerns on the Anthropocene. The ecocritical analysis of the Australian landscape through the lens of theatre and performance takes into account the distinctive features of the Australian continent, which has been affected by the history of colonisation and European settlement and the arrival of agricultural and industrial revolutions as well as ecological imperialism (Robin). The legacies of colonising practices in Australia – including environmental injustice, the dispossession of Indigenous communities, the exploitation of the land, farming and natural resource extraction, species extinction, and consumption heavy lifestyles – have considerably contributed to a global legacy left for future generations that is the emergence of a new chapter in the Earth's geological history. In this way, the Anthropocene is rooted in histories of settler colonial dominance that is environmentally catastrophic and expropriates the material bases of Indigenous ways of life. Australia's preeminent role in accelerating global emissions and, consequently, the global climate crisis and mass extinction informs the Australian contribution to Anglophone ecocritical theatre, reinforcing my reason for engaging these Australian plays in dialogue with British, American, and Canadian ones with the global quest for climate action. Australian drama scholarship has recently responded to such legacies by raising questions about our relationship with global ecological transformations and introducing a new age of Anthropocenic theatre. Although they are relatively rare, environmental plays and performances intervening in the context of climate change include Elaine Acworth's *Water Wars* (2011), Ian Meadows's *Between Two Waves* (2012), Marrukegu Theatre's interdisciplinary production *Cut the Sky* (2015), Linda Hassall's *Dust* (2015), Stephen Carleton's *The Turquoise Elephant* (2016), Mark Kilmurry's *The Big Dry* (2016), and David

Finnigan's *Kill Climate Deniers* (2016), as well as a few works produced by Doppelgangster – an international performance company currently creating work in the UK and Australia – including *Doppelgangster's Titanic* (2015), *Oxygen Support* (2015), *Baby* (2016), *Disinformation Booth* (2016), and Hanna Cormick's *Mermaid* (2020). The ecological concerns in these Australian works are not only tied to a local and easily discernible sense of time and space but also respond to ongoing planetary crises and changes.

Environmental problems and climate change are fractious and divisive political issues in Australia; in this context, I examine the theatre in this chapter. As Kate Crowley points out, "Australia is a carbon-intensive economy, with coal its largest export, and dominating electricity production" (121). Increasing carbon emissions from such great dependency on fossil fuels have made Australia not only a driving force in the geological transformation of the Earth, but also one of the nations most vulnerable to climate change. Evidence such as the 2021 wildfires, the awful 2019–2020 bushfire season and extreme weather, the "Angry Summer" of 2012–13, the 2011 floods in Queensland, the 2008–2009 Victorian bushfires, and the 2003 Canberra bushfires exemplify the reality of an impending ecological crisis in this country, calling for urgent global attention and action. In *Feminist Ecologies: Changing Environments in the Anthropocene* (2018) – which critically brings Australian ecofeminist thought and activism together with Anthropocenic environmental issues – the editors Lara Stevens, Tait, and Varney argue:

> Despite Australia's legacy of environmental activism, innovative thinking in deep ecology and formative ecofeminism it has, more recently, experienced the rise of conservative governments with strong electoral mandates that have resisted and overturned action on policies promoting sustainability and greenhouse gas reduction.
>
> (9)

An investigation of Australia's emissions control policies and Australia's position in relation to the Kyoto Protocol's targets sheds light on this country's relationship to climate change policy. Under the Kyoto Protocol, an international treaty negotiated in Japan in 1997, Australia was required to limit its greenhouse gas emissions to 8 per cent above 1990 levels by 2008–2012 (Crowley 120). However, following its attendance at the UN Framework Convention on Climate Change in 1998, Australia did not ratify the protocol in 2002. By the time, for instance, Sewell's play opened in 2006, the Kyoto Protocol was yet to be ratified under the conservative government of John Howard. It was not until December 2007 that Australia ratified the agreement under a Labor government, the first ministry of Kevin Rudd. By the return of a conservative government in 2012, however, Australia's politics of climate change took a different turn, and the country did not meet its target to reduce carbon emissions.

Pondering Australia's delay in acting on climate change policy, I examine how these two plays by Sewell and Bovell materialise the complexity of such global threats and how they contribute to the theatre of climate change. By analysing the themes of these plays within the framework of material ecocriticism, I explore how the global, non-anthropocentric phenomenon of climate change is made visible in theatre despite the limited spatial and temporal scope of the theatre. The plays are valuable models for dramatising Una Chaudhuri's critical question in "Dramaturg's notes: The Weather Changes" (2014): "How might the theatre contribute to the recognition that *Homo sapiens* is . . . in need of new ideas and new lifeways in the face of catastrophic climate change?" (122). By placing Sewell's domestic scenario and Bovell's family saga in this context, the chapter describes such change in humankind's relationship to the material world.

The theatrical works studied in both this chapter and the previous one deal with how a variety of nonhuman agents, such as toxins, weather, and energy, are entangled with human bodies in distinct ecological contexts and so reformulate human agency and ontology. In Chapter 2, my new materialist reading of Caryl Churchill's plays refers to the irreducible agency and materiality of human bodies, nonhuman animals, things, and matter, as well as to the physical world in the context of environmental pollution and global ecological crisis. The plays' narratives of matter or "storied matter" articulate trans-corporeality and the inseparability of the human from the world of vibrant matter. In this chapter, I focus, through ecocritical readings of Australian plays, on the question of scale as another central challenge in thinking about the Anthropocene. That challenge takes the new form of "expanded materialism," or matter on unthinkably large temporal and spatial scales. This new conception of materiality, with both its geographic and geologic dimensions, is founded upon an object-oriented ontological approach. Like the new materialism, this approach insists on a re-examination of dualisms such as human versus nonhuman, living versus nonliving, and subject versus object as well as a decentering of (human) subjectivity.

The plays challenge the idea of subjectivity and underscore the ecological position of humans in relation to the immense scale and distributed nature of matter that surrounds them. To show how the plays do this, I draw on Timothy Morton's concept of "hyperobjects," which represent the disturbing agentic potential of the emergent products of human and nonhuman entanglement. Morton defines hyperobjects as "things that are massively distributed in time and space relative to humans," such as global warming (*Hyperobjects* 1–2). Taking into account concepts that range from Stacy Alaimo's idea of trans-corporeality and Jane Bennett's vital materialism to Morton's hyperobjects, this chapter considers an adjacent intellectual shift in our understanding of matter and materiality. It does so by broadening our concepts of place, space, time, and matter.[3] Reading the plays through this lens also encourages us to think about the so-called end of the world by dissolving our more conventional ideas of nature, world, and worlding in a time of climate change.

1 Staging the global ecological emergency in *It Just Stopped*

Premiering at the Malthouse Theatre in Melbourne and the Belvoir St Theatre in Sydney in 2006, *It Just Stopped* has been described as "a dark and savage play" as it narrates a scenario for "the end of the world" (Elliott et al., *Drama Reloaded* 89). The play introduces us to a middle-class American couple, Beth and Franklin, who live in Melbourne on the forty-seventh floor of a high-rise luxury apartment building. It has a wall made of jellybeans, a feature that satirises the couple's contemporary taste. Franklin is a music reviewer for the *New York Review of Books* – he is writing an essay on Wagner's "Ride of the Valkyries" – while Beth produces radio programs for a shock-jock broadcaster. Their lifestyle is urban and contemporary but insulated; they have no relationship with their Australian neighbours. Their comfortable lives are interrupted by an unexpected predicament when one morning they wake to find the electrical appliances on which they depend – air-conditioning, computers, phones, and lifts as well as the internet, radio, and television – have ceased to function. This unexpected and bizarre situation, which suggests a broader ecological concern about energy-intensive lifestyles, turns out to be darker and more intense than they first imagined. The narrative descends into an ever more comic and bizarre scenario as their neighbours, Bill and his wife Pearl, arrive to announce the end of the world. The social and cultural differences between the educated, cultured American couple and the vulgar, nouveau-riche Australians become comically apparent in arguments about art and politics, until the more prosperous Bill and Pearl offer Franklin and Beth a "proposition." This proposition moves the play into a grotesque absurdity whereby the Americans become Bill and Pearl's slaves in exchange for being protected from the end of the world. With this sinister image, Act One ends. Cutting and punishing dialogue extends across Act Two as the characters engage in constant verbal conflicts. The shift from a comedy born of ideals and delusions to this absurd situation is accompanied by a scene of chaos, confusion, and horror as Franklin increasingly loses control and finally shoots Bill. Then an enormous explosion is heard as the jellybean wall collapses, and there is a horrible sound of screeching monkeys and hooting baboons, along with a frightening vision – the hanging body of the Australian couple's fat son fills the stage.

According to Mark Seton, *It Just Stopped* and Sewell's earlier play, *Myth, Propaganda and Disaster in Nazi Germany and Contemporary America* (2003), "address contemporary Western culture and its complacency and growing conservatism in the face of what he regards as 'darkening times'" (230). Sewell's concept of "darkening times" might well refer to the global ecological crisis and issues such as climate change. A reference to such an implication is evident in Robin Usher's interview with the playwright, who describes the play as "the only way to comment on the end of the world" as it is "fast, witty and funny and black, black, black" (n. pag.). This description might be related, as Usher says, to the fact that "[t]he play's background is the lack of response in Australia and the US to scientific warnings that global warming has dire implications for the planet" (n. pag.). Regarding the history of Australian climate policy and the urgency of addressing climate change, in his interview with *Run Riot* in January 2014, Sewell criticises the current toxic politics of climate change in Australia in this way:

> Over the past seven years, polling indicates that Australians have been overwhelmingly in favour of *some* form of intervention to prevent climate change (rising to 84 % in 2013), but the shift has been *away* from a sense of urgency to a more restrained, lower cost intervention (Lowy Institute polling). . . . As a result, climate change has slipped from the public agenda. . . . Of course the issue of whether or not Australia should join the world and put a price on carbon emissions is really a sideshow when it is understood that Australia is one of the largest coal exporters in the world, and consequently responsible for a major part of international carbon emissions. The real debate, therefore, must be whether or not Australia continues mining coal at all. And you can imagine, with an economy now basically geared toward the mining industry, how popular that would be.
>
> ("End of the World" n. pag.)

Indeed, the play does not address issues of climate change and carbon emissions directly. However, I argue that its scenario evokes a disturbing message about the age of ecological emergency inside of which its human characters have now awoken. The play is interested in exploring the modes of thought involved in a narrative of "dark ecology," to use Morton's term, which unsettles any fantasy about stasis or an imaginary World and Nature at the end of the world. The play's dark ecology also sets the scene for an engagement with the Anthropocene. The impact of humanity's carbon footprint is staged through the cutting off of power, which demonstrates the effect of planetary ecological change on the micro-scale of the individual human body. This story serves as a framework for a consideration of the human characters' unknowing immersion in the obvious ecological threat that is climate change; it also serves as a framework for foregrounding new ways of thinking about how humans react to one another and to nonhuman forces.

In the following sections, I examine the couples' different responses – denialism on one hand and apocalyptic warnings on the other – to the disruption of the electricity and the end of the world. I show that the play's critical approach to these responses resonates with Morton's concern; that is, how large-scale ecological changes at the dawn of the Anthropocene have already ended the "world" in the phenomenological sense, "rendering both denialism and apocalyptic environmentalism obsolete" and changing human art and experience (*Hyperobjects* 2).

Entering a new era of ecology: the impossibility of keeping a cynical distance

The play provides a satirical commentary on ignorance about the electricity generation process, without which modern life is unimaginable. From the very beginning of the play, Sewell represents the agency and urgency of energy in the form of electrical power through its sudden interruption. He also raises the question of what would happen to modern life if the fossil fuel industry went offline for an unknown duration. The predicament in which, one by one, all modern amenities suddenly cease to work creates a central narrative device, as the characters demonstrate their helplessness and dependence while struggling to maintain their cynical distance from this sudden change. Being disconnected from modern conveniences results in a mixed reaction of "panic and rationalisation" for the American couple, Franklin and Beth, "as they believe alternately that nothing is wrong and that it's the end of the world" (Sewell, *It Just Stopped* 4). This sense of uncertainty emerges from the fact that they are accustomed to their modern lifestyle and take electricity-dependent technology for granted, but "change comes unexpectedly" (3) and disturbs what Beth describes as "the equilibrium of their perfectly balanced lives" (38). Franklin is anxious that he will not be able to deliver his Wagner article for the *New York Review of Books* and Beth frets that she will not be able to get to work (see fig. 3.1).

Figure 3.1 Catherine McClements as Beth and Kim Gyngell as Franklin. *It Just Stopped* at Belvoir St Theatre, Sydney, 2006. Directed by Neil Armfield. Photo by Heidrun Löhr.

The characters face an unforeseen change that exposes how the smooth operation of modern life depends entirely upon a continuous supply of energy.

The American couple suggests that this unexpected situation might have been caused by an accident, terrorist attack, or gas poisoning. It is hard for them to determine whether life as they know it has come to an end. Franklin blurts out that he has no understanding of anything that happens in the world or how things work: "I didn't know where electricity came from or how the water got into the taps" and also "I don't know how this society works; how all the stuff that needs to get done does get done; I just assumed that it would be" (41). He refuses to vote or take part in public affairs. Beth has a similar attitude: "I work in talkback radio. . . . What do I care about the news?" (27). They argue that it is not their responsibility to be involved in public affairs or know how the world works because they just do their thing. They engage in denial as a way of dismissing the blackout as a temporary event or a "momentary interruption to service" (37). Franklin and Beth assert that they are good people; they consider themselves exempt from catastrophes, which they think only happen to people in isolated parts of the world. For example, in response to his Australian neighbour, Bill, who insists it is "the end of the world," Franklin says: "We don't want to hear these terrible things – I know they're happening to someone, someone else – in Africa or South America – but they're not happening here and they're not happening to me and I don't want to know about it! Why should I?!" (62). The sudden disruption of technology reminds the characters that electricity is like oxygen for them, but it is a product of the present system of capitalist development, which operates at the expense of both the environment and poor countries. The couple's indifference to what happens in the world brings to light a lack of "capacity for critical self-reflection" (Glow xiii). This inability to reflect leads to the childish cry: "Why me? Why us?" (Sewell, *It Just Stopped* 28). While Franklin and Beth are elitist intellectuals who "rationalise what they do in high-flown theoretical terms" and "wield cultural power," they are conservative and hypocritical, as Hilary Glow argues, and "use influential cultural positions to perpetuate self-concern and rampant individualism" rather than engage in the realities of this changing world (xi, xiii). Through them, Sewell condemns well-heeled members of the upper middle class for repudiating their social, political, ethical, and environmental responsibilities.

Although some argue that climate change is a product of global capitalism and that wealthy nations will be more able to adapt in the face of disaster, Dipesh Chakrabarty contends that "the whole story cannot be reduced to a story of capitalism" because, "[u]nlike in the crises of capitalism, there are no lifeboats here for the rich and the privileged (witness the drought in Australia or recent fires in the wealthy neighbourhoods of California)" (221). Chakrabarty reminds us instead that all humanity is trapped in a new geological era. The couple's dependence on technology and a luxurious Western lifestyle is ironically considered a kind of protection from the negative dimensions of planetary changes, which are themselves caused by excessive reliance on fossil-fuel energy. Sewell's play reveals to its audiences the central dilemma of how to escape global environmental change while remaining committed to a modern high-consumption, emissions-intensive lifestyle.

Audiences see how the American couple, and by implication themselves, forget that they naturalise everyday technological devices and fail to realise that all technology is enabled by processes of generating power that emit carbon dioxide and other greenhouse gases. The couple treats electrical appliances, the internet, and radio and TV networks as if they are not real material objects, though they contribute to the carbon footprint of affluent urbanites and their highly consumptive lifestyles. In other words, these gadgets are the traces, indexical signs, or "phasing" objects, in Morton's words, of material processes of mining coal, generating electricity, and producing carbon emissions – processes that are invisible to the American couple and often out of sight for the audience. Together they compose large objects – hyperobjects such as climate change – that are distributed globally and are outside of human control. In this case, as Morton argues, the difference between cause and sign is eliminated because climate change as a hyperobject is not experienced directly but only mediated through a widespread process of "interobjectivity." It is made apparent through other signs or objects, revealed as information or in some shared sensual space (*Hyperobjects* 86). In this mesh of interconnectivity, I argue that the sudden change that happens to the American couple is a micro-manifestation of the broader ecological change caused by anthropogenic carbon emissions. Neil Armfield, who directed the play at Belvoir Street Theatre in 2006, asserts:

> Stephen Sewell is seeking dramatic form for what he sees as the delusion we are all living: we cling to our lifestyle even though we know it is unsustainable – in fact, not only unsustainable, but actually ruinous to the world that gave us life.
>
> (qtd. in Malloch n. pag.)

Such an "unsustainable" and "ruinous" lifestyle implicitly recalls the fact that, in Australia, fossil fuel is used to generate the majority of household electricity, while the coal industry produces a high proportion of the country's total greenhouse gas emissions. Nonetheless, Sewell's characters are blind to their dangerous sense of complacency and assume that the world will go on as it always has.

Marking a change from something familiar and accessible into something unexpected and abnormal, the sudden disruption of the electricity transmitted from one object to another creates a kind of "estrangement effect" or "defamiliarisation effect," to use Bertolt Brecht's words (Brecht and Willett, *Brecht on Theatre*). This effect motivates the audience to wonder what lies hidden beneath the apparent interruption of energy. Franklin and Beth do not realise how deeply they are dependent on energy until their life ceases to function in an instant. The moment when the power and water supply and all technological devices shut down can be interpreted as an alert to Franklin and Beth that their lifestyle is not natural but dependent on social and industrial processes that are mostly hidden from view. The couple has to wrestle with the fact that their normal life has become unfamiliar. In particular, there is no exit from this disastrous situation because their selfhood is not separate from the world. In an era of global ecological change, Morton argues,

> No longer are my intimate impressions "personal" in the sense that they are "merely mine" or "subjective only." . . . I become (and so do you) a litmus test. . . . I am scooped out from the inside. My situatedness and the rhetoric of situatedness in this case is not a place of defensive self-certainty but precisely its opposite.
>
> (*Hyperobjects* 5)

It is no longer possible for Franklin and Beth – whose story is not exclusive – to maintain their distance from the ecological emergency because it has surrounded them and they have unknowingly been part of it. This way of thinking contributes to the dissolution of the distinction between inside and outside as the characters' reality is caught in material forces that act on scales vastly larger than those of the human.

However, Franklin's assertions about the purity of art and music can be seen as substantial evidence of his attempt to intensify his distance from the world. Incredibly, Franklin insists that art cannot be about anything but itself and that "there are some areas of human life that can remain out of bounds" (Sewell, *It Just Stopped* 21), thus disclosing his romantic or idealist viewpoint and his belief in a gap between human culture and the natural world. In his conversation with Bill, who has a more pragmatic and realistic understanding of art, Franklin declares: "We don't want our artists to be lecturing us about what's wrong with the world – we want to be transported to another world" (44). He describes the central theme of his article on the meaning of art in this way: "when art is about *nothing*, it suggests *everything*; but when art is about *something*, then it ceases to be about *anything* at all" (45). For instance, he claims that to interpret Wagner's musical language through the lens of social and political issues is ridiculous and that it ruins the purity of art:

> imagine what Wagner might have written about if he thought opera was about changing the world – the plight of the serf, perhaps? The women's question? Could you imagine that? Instead of the glorious march of the gods into Valhalla, we'd have the workers trudging drearily to freedom!
>
> (44)

Such a perspective on art denies the ongoing transformation of planetary life while maintaining a complacent distance from it. It is an attempt to ensure that, as Franklin says, "[t]here's nothing wrong with our world! Everything's fine" (48), and to preserve an unchangeable order in which humans are safely separated from the outside world. Such an "aesthetic distance," to use Morton's term, is key to denying the fact that the conventional idea of the natural world as a stable background existing outside the sphere of civilised human activities can no longer be sustained. This distance prevents the characters from seeing that they are enmeshed in a relationship with other living and nonliving beings.

Ecocritical thought provides a critique of Franklin's idealist concept of art. For Sewell, "in this environment art which does not speak out about politics and truth is a smokescreen," as Glow puts it, because "it allows the bourgeoisie to see itself as 'cultured' while at the same moment they turn their backs on the suffering that is taking place right under their noses" (xi). However, Franklin disagrees: for him, Wagner's music and art are not about anything. Franklin describes art as a holy place that can protect us from changes that happen in the world by distracting and amusing us, making us indifferent to them, delaying ethical and political decisions until the advent of an ideal future, and keeping us at a distance from the world. According to Morton, modernity, due to its anthropocentrism, defined art via the distance between the natural world and the human. However, such an aesthetic no longer works in the age of hyperobjects, which is directly affected and "shaped by the current ecological emergency" ("Art in the Age of Asymmetry" 122) that is beyond the scope of our imagination.[4] As Sewell states, "the world itself intervenes and Beth and Franklin discover to their rising horror that all the verities of their lives, even down to the availability of food and water, are in the process of mysteriously disappearing" (*It Just Stopped* 4). Indeed, as I argue, Sewell's perspective is grounded in what we might identify as ecological thought. His approach resonates with Morton's idea of "ecology without Nature" in which he argues that the concept of Nature – as a holism – impedes ecological thinking by creating a barrier between inside and outside (*Ecological Thought* 9). Truly ecological thinking goes beyond the conventional distance between the human and nonhuman environment to call forth an awareness of coexistence rather than detachment. Morton argues that during the Anthropocene epoch, we live inside massive objects such as climate change, which are the primary cause of "the end of the world" (*Hyperobjects* 2), have their own agency, and are able "to make decisive contact with humans" (5). The massiveness of these objects, along with our inability to measure or understand them in the restricted scale of our time and space, undoes the abstract ideas of "world and worlding" (103), which are based on a distinction between humans and nonhumans. Hence, the end of the world is, for Morton, the end of the concepts of World and Nature and "all the modern certainties" (130). In the context of our coexistence with hyperobjects, no entities exist beyond interconnected relationships, and there is no predictable conclusion and no vessel, such as World and Nature, in which all life and non-life forms float and are unified (128–9). This way of understanding the end of the world compels us to reconsider how we are connected to the ongoing changes that are already occurring in the Earth's ecosystems. The play underlines how the human characters and their lives are deeply affected by the sudden change of the power outage. Reading the play through this perspective opens up the possibility of dark ecology by asking us to think big and include darkness and negativity as well as lightness and positivity in our thinking about the physical world, the interaction among humans and nonhumans, and the meaning of art.

Apocalypse no?

In this play, the panic that follows the removal of modern amenities is a catalyst for further violent, uncivil, and unscientific responses. The "unexpected arrival" of Bill and Pearl moves the play into "darker territory" (Varney, *Radical Visions* 263) that includes slavery, brutality, and murder. Bill, the cardboard businessman, recalls apocalyptic movies such as *The Day after Tomorrow* and *The Book of Revelation*, and makes proclamations about "The End, the Apocalypse, the Last Day, Judgment Day, Armageddon, the Number of the Beast, and the war between Gog and Magog" (Sewell, *It Just Stopped* 34), seeking biblical explanations for the dramatic turn of events.[5] He then shifts register to the comic and colloquial, saying "it's bloody obvious," "[e]verything's been going to the shit for so long the whole thing's just given up the ghost and collapsed," and "[t]his is the end . . . it is no more" (37). Bill sees a business opportunity if he can convince Franklin and Beth that this interruption is not just temporary (see fig. 3.2). Beth accuses Franklin of useless idealism now that time has run out. Franklin, however, thinks nothing has changed and suggests Beth should also think everything is fine because they'll be much happier that way.

The Australian couple tries to take advantage of the situation by overemphasising apocalyptic prophecy and doom. As the playwright says in an interview, Bill and Pearl are "the strangely demonic tormentors of the ruling elite, a kind of *return of the repressed*, who stir the terror of the first class passengers as they watch their privileged existence plunge toward its fatal and all too predictable end" ("End of the World" n. pag.). Here, the playwright reflects on the scenario from an Australian point of view that sees American elites as the ruling class of a declining power. Though the cynical Bill does not deny the end of the world like Franklin does and insists on telling it like it is, his apocalyptic language postpones the end and embraces a hypothetical future. Bill's language creates, paradoxically, a distance from the real change (or hyperobject) that has encroached on their lives, and replaces it with a grotesque future. Bill's statement about the end of the world fails to find a solution to the current problem because it sensationalises and overstates events and generates further violence and crisis as the play moves towards its end. Further, Bill's apocalyptic rhetoric is problematic because it depicts the end abstractly and inhibits intimacy with the material world. For example, in response to Beth's question that seeks "a perfectly reasonable, rational explanation" for the power outage (Sewell, *It Just Stopped* 37), Bill says: "Who cares how it happened?" (38). Morton asserts that such apocalyptic language is based on "a fundamental lack of concern for the way things are going" and the belief that "[s]ince the end of the world is nigh, or since we will all become extinct in the long run, there isn't much point in caring" (*Ecological Thought* 27). The emergence of this type of apocalyptic account is troubling because, as Rigby argues, it may "engender either a grim

Figure 3.2 John Wood as Bill and Rebecca Massey as Pearl. *It Just Stopped* at Belvoir St Theatre, Sydney, 2006. Directed by Neil Armfield. Photo by Heidrun Löhr.

eagerness for the end that . . . heralds a glorious new beginning, or an enervating despair, potentially combined with a heedless hedonism: 'Since we're all doomed anyway, let's live it up while we can'" ("Imagining Catastrophe" 70). Bill's narrative resonates with the latter because he provokes a despairing view of the world situation in order to fulfil his plans. Still, Morton argues that such eco-apocalyptic thought "warns of, and wards off, the end of the world" (98). While Bill "warns of" the end of the world, he also "wards off" its reality by preventing any intimacy with the world and its ongoing change, and he fails to take responsibility for it.

The play reveals two different reactions when the electricity stops working, and these reactions satirically suggest how dependent the characters are on energy and how difficult it is for them to deal with reality, to cope with change, and to adapt. The couples' responses resonate with obsolete attitudes towards global climate change, including what Morton would describe as "denialism" and the fatalistic essence of "apocalyptic environmentalism" (*Hyperobjects* 2). The play critiques Bill's thinking and represents him as monstrous, but it is equally critical of the elite. As I explained earlier, in the play the strategy of denialism or self-deceptive optimism is adopted by the educated, cultured middle-class couple, Franklin and Beth, who believe in the stability of the physical world as a background to their lives and irresponsibly deny the dark reality of planetary change. However, in the current era of increasingly frequent and ferocious Australian environmental disasters, it is impossible to achieve an absolute distance from the ongoing changes taking place in the physical world. The Australian couple takes a different tack: they invoke the cynical language of apocalypse that, according to Glow, is the "rat cunning" of "a bully who relishes in and excels at the production and dissemination of fear" (xiv). Drawing on a pessimistic strain of apocalyptic thinking, Bill and Pearl assure the American couple that it is truly the end of the world, though their real concern is not to save the world from more destruction or to contemplate why and how it happened, but to profit from a disastrous situation (see fig. 3.3), as the following exchange demonstrates:

BILL: We've got to face facts, Beth; it's not looking too good, is it?
BETH: But how? How?
BILL: Who knows how? There could be a hundred and one ways how. In fact there's so many ways you wonder how we even got as far as we did, but now it's over and we just have to make the best of things. (Sewell, *It Just Stopped* 40)

Even in response to Franklin's question about how they can save themselves at the end of the world, Bill says that people cannot save themselves and then states his own intention very explicitly: "Human beings are too mad . . . and too selfish to do anything other than look after their most immediate interests, and if those immediate interests mean goodbye Amazon to pay for the kid's boarding school in Pennsylvania, then goodbye Amazon" (69). What Bill and Pearl know is that the end of the world has "certain inescapable implications" (Sewell, *It Just Stopped* 38), and they need to avert the dreadful consequences by implementing their own business plan to guarantee their safety. Using "coercive, bully-boy tactics" in "a social Darwinist game" (Glow xiv), they make their weak neighbours, Franklin and Beth, a "business proposal" whereby the American couple becomes their slaves. This creates an unexpected moment in which everything takes an absurdist turn, as the play moves to Act Two.

Figure 3.3 Catherine McClements as Beth and Kim Gyngell as Franklin. *It Just Stopped* at Belvoir St Theatre, Sydney, 2006. Directed by Neil Armfield. Photo by Heidrun Löhr.

Dealing with anti-scientific thinking in the Anthropocene

The strange and sinister events of Act Two offer opportunities to think about the question of what happens when societies repudiate the scientific view of the world. In this section, I thus shift my discussion about the end of the world to another context. Now that everything on which they depend has stopped functioning, how is it that the couples' differing responses to their situation lead them back to the idea of slavery? As the play continues, Franklin insists on his belief that he and Beth are civilised and enlightened, and so immune from harm: "What if we survived so far because of who we are?" (Sewell, *It Just Stopped* 52). While the couple needs to yield to "a panic-driven re-assessment of their careers, lifestyle and politics" (Varney, *Radical Visions* 264), Franklin cannot accept that they must change now that everything has changed. He is trapped in the delusory projection of a supposedly secure and unchangeable individual life despite the terrible disaster that has taken place. By repudiating the fact that he and Beth are undergoing the consequences of environmental crisis, he unknowingly prepares the stage for the exploitative proposal of their more prosperous and acquisitive Australian neighbours. In exchange for their slavery, the wealthy couple will offer Franklin and Beth a "key to life" – that is, a key to a state-of-the-art helicopter that will keep them safe from chaos for a lifetime, as the world's resources disappear. It can take them to "a rather sweet little island protected by a very clever sea wall against the rising waters and hidden somewhere in the southern ocean" (Sewell, *It Just Stopped* 55). In return, Bill says that he and Pearl need them to "keep our minds off things" and entertain them (53). Franklin is an intellectual who can entertain them by giving lectures on pure art and dissociating Wagner's music from anti-Semitism, while Beth can amuse them by acting out shock-jock programs with their melodramatic exaggerations. In response to Franklin, who asks what their rights will be, Bill responds: "I don't know if you've paid a lot of attention to human history, Franklin, but I'd have to say that slaves don't have an awful lot of rights" (58). Bill also reveals that as slaves, Franklin and Beth should bear children so that they can be grilled and eaten. The horror and absurdity of the situation make Beth wonder about the kind of world opening before them. These surreal and grotesque events also satirically expose the anti-scientific presumptions of the intellectual couple.

In a broader sense, what makes the insanity of the situation so much more dismaying is the characters' inability to acknowledge the agency and materiality of environmental threats themselves. What is actually needed here is a rational and scientific response to anthropogenic carbon emissions – caused in part by the generation of electricity from fossil fuels – and their ecological effects. Indeed, in order to acknowledge the inextricable connection between global environmental change and everyday life, the characters need "the Enlightenment (that is, reason) even more than in the past" (Chakrabarty 211). However, the couples are the captives of their idealism, romanticism, and selfishness, as well as their modern, affluent, and, in fact, environmentally irresponsible lifestyle. In other words, audiences can see that the characters are ironically incarcerated in their own disastrously increasing energy consumption and its environmental impact. Also, the characters show vulnerability in their inability to take control of the cumulative results of their oblivion. They are not aware of how much their decisions and freedom are defined by an inevitable entanglement with the consequences of their unsustainable lifestyle.

The play's bizarre events do not stop at the slavery proposal; rather, they proceed to more complicated and tragic stories, which are accompanied by the materiality of one formal element: the play's sound. Driven by Wagnerian ideals and a hysterical attachment to the aestheticization of life, art, and objects, Franklin shoots Bill. As Seton points out, "Franklin's refusal to relinquish the arts as proof that humanity is worthy of its ascendency above the rest of the created order leads him into a kind of ecstatic Dionysian frenzy" (232). That bizarre moment is accompanied by the reverberation of contradictory sounds around the theatre: the sound of "Ride of the Valkyries" rising from the stairwell, the sound of doors breaking and fire crackling outside, the sound of an explosion from the collapse of the jellybean wall, and the horrible sound of screeching monkeys and hooting baboons as the hanging body of Bill and Pearl's son swings into view. Everything falls apart and the characters' modern life descends into mayhem; meanwhile, Franklin declares that he is at the end of history as he hears the sound of glory and destiny, enters Valhalla, and achieves "Nothing" that is "the last conqueror of *all!*" (Sewell, *It Just Stopped* 73). In the final moments of the play, "*we see* FRANKLIN *at the end of time, the last human being on earth, his face transfigured with joy*" at the same time as "*the baboons cry in fear and fury*" (73). In the last moments of the play, these shrieks rise over the sound of Wagner's music, revealing how destructive the mode of reality that Franklin takes as a norm is. The play shows how little Wagner has to offer human society in its moments of doubt and fear. I interpret Franklin's assertion that he stands at the end of history, as the last human being on earth, as the beginning of a new phase of history that is not exclusively human. According to Morton,

> the end of the world has already happened. We sprayed the DDT. We exploded the nuclear bombs. We changed the climate. This is what it looks like after the end of the world. Today is not the end of history. We're living at the beginning of history. The ecological thought thinks forward. It knows that we have only just begun, like someone waking up from a dream.
>
> (*Ecological Thought* 98)

This mode of ecological thinking provides a corrective to Franklin's illusion of a salvific ending and exemption from the changing world. Morton's definition of the end of the world might be understood as a future which has, in a sense, already arrived. We are living with the Anthropocene. Today is a time of environmental crisis, which we have caused and with which we are now entangled. In response to the present context of ecological change, the play's ending whimsically and satirically announces broader realities: the end of the world of the ideal and metaphysical imagination and the beginning of a dark ecological world evoking the interconnectedness of everything. The play is haunted by ominous and complex realities that resist full comprehension and so are denied. The couples are involved in a comic, absurd world and are blind to the dark realities of their environment. They do not know that it is only by collapsing the distance between the human world and the material world that they can hear the sound of the breakdown of the human dream. They mistakenly believe that if they keep this distance in place they can remain safe in the cocoon of unitary individuality and self-complacency supported by their modern lifestyle. The play integrates this dark reality into the materiality of the sound, which emerges as a powerful cue, contributing to the play's formal imagination of "expanded materialism." In other words, the sounds are traces of a massive change: a hyperobject has been embedded profoundly in the characters' lives and invites us to think of the end of the world.

The play derives dramatic force from the challenging idea that, if the end of the world has already happened, a non-anthropocentric narrative of the world and a new concept of humanity must emerge to reshape our self-conception and conventional ideas of self-interest. Sewell calls for this new humanity by countering those who insist on their delusive separation from the rest of the world and act absurdly in response to sudden changes in their daily lifestyle. In effect, the play provokes us to think and act ecologically, ethically, and decisively in the face of the unexpected potential effects of climate disruption, which reveal themselves in the familiar life-world of everyday human experience on an individual and personal scale.

2 Traces of the hyperobject in *When the Rain Stops Falling*

If, in Sewell's play, the sudden disruption of electricity underscores the materiality of energy and serves as a potent indexical sign of the hyperobject of climate change, in Bovell's *When the Rain Stops Falling*, the agency of weather is another trace of the changing climate represented in its local manifestation in the Australian context. In this part of the chapter, I analyse this 2009 Australian theatrical work from a material ecocritical perspective in the context of climate change. In taking this theoretical approach to the play, I focus on the key concepts of the end of the world, climate, and time and space. Reading the play through material ecocriticism opens up questions about how theatre can narrate the materiality of a vast entity, such as climate in this play, through interaction with the human drama. Sewell's play is a revealing critique of the relationship between an energy-intensive, high-emissions modern lifestyle and individual human experience. This kind of lifestyle is a driving force in destabilising the climate and damaging ecosystems. Bovell narrates such environmental disturbance in a different way by showing how it intrudes on an intricate web of connections between generations. I will tease out the challenge of dealing with and grasping the large-scale phenomenon of climate change in theatrical space by examining how Bovell's play gives dramatic form to a rethinking of the relationship between people across time and space and their interaction with nonhuman forces. As the play turns to environmental concerns, its capacity for making global ecological interconnections allows us to consider the following questions: how do the planetary symptoms of the Anthropocene affect "human social and psychic space," emotion, perception, ethics, and art? (Morton, *Hyperobjects* 2). And how does the spatiality of theatrical performance shape ecological thought, connection, and community? Focusing on these inquiries illuminates how the play engages with climate change and related risks in Australia. This contextual reading allows for a broader exploration of how relationships among humans and between humans and nonhumans are affected by global ecological threats.

The play unfolds non-chronologically and features interconnected events in the lives of four generations of one family, traversing time and space over a period of eighty years, from 1959 to 2039, and between London and a variety of places in Australia – Alice Springs, Adelaide, the Coorong, and Uluru. The stage directions in the first scene evoke magical realism, describing a theatrical landscape of ecological catastrophe in 2039 as the rain falls relentlessly and a fish that was considered extinct falls from the sky at the feet of Gabriel York in

Alice Springs (see fig. 3.4). This strange scene was predicted seventy years earlier, around 1970, in a series of enigmatic postcards that his grandfather, Henry Law, sent from Australia. Henry had a vision of the end of the world: "Dear Son, in the desert I saw a vision of the end. A fish fell from the sky and the earth became sea" (Bovell, *Rain Stops Falling* 45). The play's scenes move back and forth between the worlds of these two men, eventually revealing a terrible crime that took place in Henry's time. The story of Henry's son Gabriel Law and his girlfriend Gabrielle York moves between the Coorong and Uluru, in southern and central Australia respectively, as Gabriel, in 1988, attempts to retrace his father's footsteps to solve the mystery of his disappearance and find out why he left the family in London when Gabriel himself was very young.

Figure 3.4 Jason Whyte as Gabriel York. *When the Rain Stops Falling*. Circa Theatre, Wellington, New Zealand, 2011. Directed by Susan Wilson. Photo by Stephen A'Court.

Finally, while discovering the terrible truth that his father committed paedophilic acts and killed a child in the remote Coorong, Gabriel learns that the victim was Gabrielle's brother. Gabriel Law dies in a car accident while Gabrielle, pregnant at the time of the accident, names her boy Gabriel York after his father. Later she becomes estranged from her adult son. The play begins and ends in 2039, with Gabriel York waiting for his son Andrew Price to arrive for lunch, which will be the extinct fish that fell from the sky. Between them is a line of dead ancestors who pass on family relics to Andrew. All four generations eat fish, and the rain stops falling.

The play is a collaborative work by playwright Bovell, visual artist Hossein Valamanesh, musician and composer Quincy Grant, and Chris Drummond, the director of the world premiere season in 2008. Looking back at the play's ecological context, Drummond recalls that it was inspired by the bleak and despairing situation of global warming. The play was originally titled *The Extinction Project* and was influenced by Tim Flannery's *The Future Eaters* (1994), which is about the ecological history of Australia and the disequilibrium of the world's ecosystems as a result of human-induced environmental degradation and resource depletion. According to Drummond, the book contributed to thinking about "the way we live our lives" and to exploring "humanity's relationship to the planet, to the unknown and to one another" (viii). In his recent interview with Walter Bilderback, Bovell refers to the importance of Flannery's concern that we are "destroying our own habitat to the extent that we could cause our own extinction" ("Zeitgeist" n. pag.). Influenced by Flannery's idea of "consuming our future," the playwright explains that extinction was the thematic starting point for the play, which "naturally led to a conversation around global warming and the potential ramifications of a changing climate" (qtd. in Bramwell xiv). For Bovell, this Anthropocenic concern was the basis of the whole investigation. In addition, the play's time setting and overlapping connections between the past, present, and future arise from its ecological interest. For example, regarding the importance of placing the Law family saga in the context of the "Great Acceleration" in human activity and carbon emissions – with its consequences for the Earth system – Bovell asserts that "the time setting was intentional" ("Zeitgeist" n. pag.). He adds:

> Originally, Henry Law was a climate scientist engaged with the study of the weather. Reference by scientists was first made to "man made" climate change in the 1950s. I took this early notion of the idea as a starting point. But I also wanted our own time . . . roughly 2013 to sit at the centre of the work and for the past and the future to radiate out from there. Our time being the focus for the debate around climate change.
>
> (n. pag.)

In spite of the ecological framing of the play, the conformity of the temporal setting with this frame, and its narration of climate change as a driving force of extinction, many critical interpretations approach the play as a human-centred family drama rather than an ecological one. This might be because the play's allusions to extreme weather, the extinct fish, rain in Australia, and flooding in Bangladesh – which are local exemplifications of a planetary change – are considered only as scenic backdrops to, or metaphors for human interests in, a family saga stretching from 1959 to 2039. By discussing the idea of climate change as a hyperobject that is embedded in theatrical space, my reading of the play underlines how it decenters the human. My reading approaches the play from an ecological perspective, showing how human actions and family relationships are shaped by global environmental change.

Weather vs. climate change

The play's representation of climate change conveys the growing sense that it is no longer possible to have a stable and predictable conception of weather or even to have a routine conversation about it.[6] As the play begins, a variety of conversations refer to people drowning due to floods on other continents, such as in East Pakistan, which the text later calls Bangladesh, a low-lying region on a delta in South Asia. With increasing levels of poverty and population density, Bangladesh is prone to famine, floods, rising sea levels, and extreme weather (Mostafa and Bose). As one of the most precarious places in the world (Mostafa and Bose), this country acts as a signifier of the imminent threat of global ecological catastrophe in this play. A reiterated assertion about the incessant rain in the play is that "there are people drowning in Bangladesh, so we shouldn't complain." While it reappears as a leitmotif through the play, repeated by characters of three generations, this assertion turns out to be meaningful by the final generation, in 2039, as a reference to a global phenomenon. In Alice Springs in central Australia in 2039, Andrew reports that a real "end of the world" event is occurring in Bangladesh and spreading to other parts of the world:

ANDREW: I'm sorry I'm late. I was held up by the rain.

GABRIEL: Terrible weather. In Alice Springs! Still, there are people drowning in Bangladesh, so we shouldn't complain.

ANDREW: I know. It's terrible.

GABRIEL: What?

ANDREW: What's happening in Bangladesh. [*Beat.*] You haven't heard the news.

GABRIEL: No, I don't follow it.

ANDREW: There's unprecedented flooding. Large parts of the country are under water. The death toll is near half a million and rising. And it's not just Bangladesh. Severe flooding is threatening low-lying areas of Northern Europe and Southern America . . . There are people out there saying this is the end.

GABRIEL: The end of what?

ANDREW: The end of the world . . . (Bovell, *Rain Stops Falling* 53–4)

The play proposes that the impacts of changing climate are not limited to particular areas of the world because events in one place can have far-reaching, disastrous effects somewhere else. This view of the end of the world, as Andrew describes, recalls Timothy Clark's question that "[w]here, especially in the Anthropocene, does 'my environment' end? Someone living a high-carbon lifestyle in New York or the Scottish Highlands is already lurking as a destructive interloper on the floodplains of Bangladesh" ("Phenomenology" 288). While Andrew's insulated father Gabriel is ironically reading a history book called *The Decline and Fall of the American Empire, 1975–2015*, Andrew tells him that events that embody the end of the world have already occurred and a changing climate is flooding coastlines around the world. This way of looking at an anthropogenic environmental crisis differs from the usual apocalyptic writing that postpones doomsday to some date in the future in order to "inoculate us against the very real object that has intruded into ecological, social, and psychic space" (Morton, *Hyperobjects* 103–4). In the play, terrible weather and torrential rains occur in every scene, and each weather event is set in a different time and place as local manifestations of anthropogenic climate change that is already happening and driving environmental transformation on a scale and at a pace that exceeds our comprehension.

The play shows what happens when climate change enters the stage and how extreme weather, as a trace of this hyperobject, embeds itself in the play's detail, extending beyond the characters' control and changing their ideas about the world and nature. In order to understand how the play gives the large-scale phenomenon of climate change a human dimension, I focus on Andrew's great-grandfather Henry in 1959 and consider his dark and problematic legacy. Near the beginning of the play, as it moves back to 1959 when the first scientific references were made to the impact of anthropogenic carbon emissions on the global climate (Bovell, "Zeitgeist"), we see Henry's interest in the history of weather. He is fascinated by the eruption of the Mount Tambora volcano in 1815 on the island of Sumbawa in Indonesia, which caused a volcanic winter along with famine and crop failures in different parts of the planet, including North America, parts of Europe and Asia:

> HENRY: Two hundred thousand dead in Europe alone. Uncounted more in Asia and Russia. And in America hundreds of thousands flee west trying to escape the weather. It shifted whole populations. . . . It changed the course of history. The weather! And it made me think just how helpless we are when the weather turns against us. And all our science. All our knowledge. All our magnificent endeavor amounts to very little in the face of bad weather. (Bovell, *Rain Stops Falling* 11)

While the volcano's effects are not the same as those of climate change, and in 1815 the greenhouse gas emissions that cause climate change had only just begun, the play evokes climate change through Henry's obsession with the 1815 volcano. The weather Henry describes is weird weather and might be considered as another depiction of calamitous weather extremes such as unprecedented flooding in Bangladesh and low-lying areas of Northern Europe and Southern America – as Andrew in 2039 describes. Throughout the play, the problem of weather is foregrounded by hearing the sound of rain falling constantly and referring to different historical catastrophic weather events, warning that "[t]he weather is turning against us," as "[i]t should not be snowing in Adelaide" (Bovell, *Rain Stops Falling* 46) – as Gabriel York's step-father Joe Ryan says in 2013. The play's insistence on the problem of weather, specifically, calls forth the end of the world and, in a broader sense, a fundamental shift in earth history and human history induced by past and current anthropogenic activities. Seen in this light, the play disturbs the traditional view of history as an exclusively human affair because the emergence of the Anthropocene age and the risk of climate change and mass extinction mark the convergence of human history and geological history.

While Henry is obsessed with the dark reality of the end of the world that is presaged by the explosion of Mount Tambora and the subsequent "Year Without a Summer" in 1816, his wife Elizabeth expresses great passion about the Enlightenment and thinks that the very fact that 1817 followed 1816 showed human resilience in the face of crisis. She also recalls that Mary Shelley wrote *Frankenstein* in the unusually long and bad winter of 1816, saying: "It will take more than bad weather to silence the human mind" (Bovell, *Rain Stops Falling* 11). The opposition between catastrophic weather events and the resilience of the human mind plays out again when Henry dwells on the deadly outcomes of a hurricane in the Caribbean and Elizabeth celebrates Denis Diderot finishing his *Encyclopedia* in Paris. Via the opposing attitudes of the married couple, the play links ecological catastrophe in one part of the world to the art and language of humans in another. For Bovell, there is "a really interesting juxtaposition between extreme weather and creative endeavor. While one is an impulse of great destruction, the other is one of great creativity" (Gough-Brady n. pag.). The playwright underlines the possibility of deep thinking evolving out of times of catastrophe and shows how the play's representation of extreme weather and flooding, as well as the rift in the family, necessitates a shift in the way we imagine and understand the material world.

Nevertheless, extreme weather plays on Henry's mind. He is not just fascinated by extreme weather such as volcanoes and hurricanes, but, as the Adelaide-based critic Lachlan Colquhoun suggests, his inner psychological tumult "is expressed in terms of weather" (n. pag.). His inappropriate acts – masturbating on a crowded train, touching a seven-year-old boy in a public lavatory, and hiding a collection of child pornography, including pictures of his son – are expressed in terms of weird weather, hurricanes, and storms. For example, when Elizabeth asks Henry what he was thinking about when he was masturbating on the train, he replies the weather, specifically, the 1780 Hurricane Callixtus which tore the Caribbean apart (Bovell, *Rain Stops Falliing* 25). The play situates Henry in a world approaching catastrophe. The theme of paedophilia and the way Henry represents a dark version of nature evoke "how we as a species have treated the planet since industrialization" (Bovell, *Rain Stops Falling* xv) and "how our actions, like weather, sometimes driven by forces out of our control" (Payne et al., n. pag.). The play's language of trauma – its repetition and generational effects – injects environmental damage and disturbance with the terror of child abuse. Through this problematic and complex parallelism, the psychological trauma evident in Henry's actions – which shatter successive generations of one family spread across two continents, with pieces of their history and identity missing from their lives – embodies human harm to the Earth's ecosystems and their deviation from balance and naturalness to strangeness and weirdness. Henry's distressed mind and unnatural act of devouring, hurting, and murdering the young boys on whom he preys might dramatise a shift from nature as pure, or as reliable backdrop, to nature as menacing and monstrous. This shift is also represented through his obsession with extreme weather and melancholic refusal to accede to his wife's romantic thinking about the natural world, the evolution of the human mind, and the ability of man to save himself.

Henry's troubled legacy also has a pervasive and persistent effect on future generations and family relationships. In 1988, Henry's son Gabriel Law retraces his father's footsteps, starting in the Coorong on the windswept coast of South Australia, in an attempt to solve the mystery of his disappearance. In a letter he wrote to his son from Ayres Rock or Uluru in Central Australia in 1970, Henry predicts the end of the world, in which "fish fell from the sky and the earth became sea," while referring to the Greek myth of "Saturn Devouring His Children" (Bovell, *Rain Stops Falling* 45). He tells Gabriel that the word *planet* means wanderer and that Saturn is named after the Roman god who devoured his own son. This story illustrates how Henry's paedophilia has damaged their father-son relationship. We can see this trauma played out across future generations, as in the problematic relationship between Henry's grandson Gabriel York and his son Andrew. In Alice Springs in 2039, Gabriel lives in the middle of a desert with relentless weather and continuous rain, suffering from loneliness, miscommunication, and a distraught mind. He has not seen his son Andrew for twenty years, although he is an attentive father and feels guilty for having left him as a child. For Bovell, the idea of abusing and devouring future generations brings him back to Flannery's book and *The Extinction Project*, which raises the question of how our past activities have exhausted planetary resources and unsettled the future (Bovell, *Rain Stops Falling* xv). Flannery's book informs the play's representation of a dark ecology embodied in Henry's feelings of shame, abjection, bleakness, inhumanity, mystery and loss, and leading to an uncertain and corrupted future.

While the shattering of these father-son relationships across four generations is not affected directly by climate change, the weather acts as a constant framing device that interacts with and underscores the human drama. The constant rain delivers "the doom of the hyperobject" (Morton, *Hyperobjects*

149) in an uncertain and unknown future in which even unborn generations are imprisoned. Henry's consciousness of how weather changes the course of history and how it turns against us resonates with Chaudhuri's announcement that

> For the first time in human history, the weather is about us. So now, when the weather changes, shouldn't we? Aren't we? Haven't we already? What is it like to be the climate change generation – the ones who have to admit that it's happening now, not later, and it's happening to us, not just to others elsewhere? What emotions . . . speak to this ecological reality?
>
> ("Dramaturg's notes" 121–2)

The way Chaudhuri describes the changing weather of the Anthropocene is a thoughtful response to the crucial question of to what extent the play's family saga parallels the reality of human geological agency. In a story of internal conflicts in generations of a family that span the past, present, and future of climate change, the playwright uses the drama of the end of the world to create a correspondence between the history of human impacts on ecosystems and how the future comes to be. In other words, two central maelstroms – one in the environment and the other in the human characters' relationships – happen concurrently and affect each other, revealing that human actions in the past are transmitted to the future, shaping not only the human social realm but also the physical world. According to the playwright, "the play is itself a description of the relationship between a family saga and the Anthropocene" (Bovell, "Zeitgeist" n. pag.). This relationship is described in terms of the Henry's legacy across three generations and the play's narrative structure over a period of time. Implicitly, these suggest an alternative to the legacy of our fossil-fuel-based lifestyles that has revealed itself in a new geological era of climate change and mass extinction. Bovell asserts that "[w]e inherit what is unresolved from the past and if we do not resolve it ourselves we pass it on to our descendants. With each generation the wound becomes deeper" (n. pag.). Gabriel, for example, declares that "whilst I know a great deal about the decline and fall of the American Empire, my own past escapes me. . . . The past is a mystery" (Bovell, *Rain Stops Falling* 5). Of course, in 2039 Andrew finds the courage to face and defy the cycle of legacy, denial, and distance – in a way that his father Gabriel does not – by beginning a new search for his family's past, reconciling and healing the fractured father–son relationship, and pursuing intimacy and love. In a broader sense, each of us – as members of "the climate change generation" – has to face the legacy of irresponsibility and neglect of the planet that we are bequeathing to present and future generations of all living creatures. Therefore, the play's potential to respond to Chaudhuri's questions is revealed "in the human experience, within the relationship between people, defined as they always are by notions of love, connection and betrayal" (Bovell, "Zeitgeist" n. pag.), in order to make climate change more palpable and moving.[7]

Zero landscape: challenging spatial and temporal boundaries

In this final section of the chapter, I consider the unbounded locality of the Coorong as described in Henry's postcard and the theatre's flexible use of time and space. The play's expanded spatial and temporal framework shows how climate change as a material force vastly distributed in time and space can be captured through the medium of theatre. The play's constant movements back and forth in time and place convert the temporal and spatial dispersion of environmental change into an interconnected web of images and narratives.

The play's local audiences know that its geographical contexts, the Coorong and Uluru, are critical environmental spaces. The Coorong and the Murray-Darling Basin in South Australia have fragile ecosystems and have been severely affected by farming and agricultural practices as well as climate change (Lindenmayer).[8] A collaborative research project conducted by Water for a Healthy Country, the flagship study of the ecological condition of the Coorong, Lower Lakes, and Murray Mouth (CLLAMM), shows that climate change has resulted in damage to these areas, including reduced water flows, increased evaporation rates and salinity, along with the loss of livelihoods, birdlife, and most sea creatures along the coast of this area (Lester and Fairweather, "Ecosystem Response Modelling"). In this way, the landscape of the play does not represent a utopian vision of capital-N nature in a traditional understanding of ecology but something dark, obscure, and weird.

Henry describes the Coorong as a dangerous place in his 1970 postcards to his son Gabriel. He looks across the darkness of the landscape and is fascinated by its uncertainty:

> Caught between the land and the sea it belongs to neither. . . . If you look across the desert, the earth takes on the appearance of the sea. You think you're standing upon a rock that rises from solid ground only to discover that you're standing on an island in the middle of the ocean. And you don't know if you're looking back into the past or into the future. . . . Water covered this earth and water will cover it again and the days that man walked here will prove just a moment in time. All our knowledge, all our science, all our money and all our will won't stop it. It's too late. . . . And time will go on without us and it will be as if we were never here.
>
> (Bovell, *Rain Stops Falling* 45–6)

Henry recounts land changing into water and water into land, as well as the disruption of temporal and spatial boundaries. Henry's paedophilic act and killing of a child in this uncertain place, and the impact this damaged past has on future generations, might refer to global change across deep time and how our past informs the future. Henry's crime might echo the way human activities have altered the planet to such a degree that our carbon footprint will still be visible in the fossil record and in the geological strata in millions of years' time. In this expanded temporal context, "you don't know if you're looking back into the past or into the future" (Bovell, *Rain Stops Falling* 45), as Henry writes. His Coorong letter depicts a micro-manifestation of phenomena of much greater magnitude, whose physical, spatial, and temporal scale disturbs anthropocentric notions of time and place. Henry's depiction contributes to our understanding of two main features of the hyperobject of climate change, "temporal undulation" and "nonlocality." As Morton says, "hyperobjects themselves prevent us from being presentist. The present is precisely nowhere to be found in the yawning Rift opening between the future and past, essence and appearance" (*Hyperobjects* 92). The play's dramatic frame reflects this feature of climate change: the play crosses temporal boundaries with no solid and fixed sense of present time and no definite distinction between the past, present, and future. Also, in the age of climate change, no place and no part of our lives will remain unaffected. In his essay "Zero Landscape in the Time of Hyperobjects," Morton argues that "[n]o landscape is big enough or long lasting enough to enclose hyperobjects in its frame" (83). In the play, the hyperobject traverses boundaries, and the Coorong is no longer a distinct landscape. Displacement – or the uncertainty of the landscape in the play – is one of the primary performative elements that arises from climate change and leads to a redefinition of the concept of home with its traditional meaning of belonging and stability.

In delineating the human condition in the face of such temporal and spatial uncertainty, Henry's letter calls for a re-envisioning of humanism and the human-nonhuman relationship. According to Morton, there is no distance and fixed frame separating humans from the landscape or the viewer from the viewed. He asserts that "[t]o have a truly ecological view we must exit from this idea of landscape, based on a first- or third-person perspective, and instead look for a *zero-person perspective*" ("Zero Landscapes" 80). The play's human characters have shattered and fragmented identities that reach across local boundaries. Each generation belongs to a specific local place and time. However, the play narrates material, territorial, and ecological connections in a complex "eco-global" network shaped by the legacy each generation leaves to the next until nothing can stop the changing weather and entry into the age of humans. Henry's narration of the uncertainty of the Coorong foreshadows the current predicament of climate change, which decentres human subjectivity and free will as well as risks our imminent extinction as a species.

The theatrical space of the play is an attempt to demonstrate both the intimacy of, and the gaps in, the world's ecological interconnectedness. The boundaries between different times and places are rendered indistinguishable by coincidences and the repetition of many expressions, events, and experiences shared by people from different generations. These shared experiences create intimacy and connection between the characters and the hyperobject of climate change. This intimacy is also due to the representational impact of extreme weather in places like Bangladesh, which are portrayed as far away but nevertheless affect the characters' familiar ways of living and thinking in a different part of the world. As the 1815 volcanic eruption in Indonesia left its traces in Europe and America, the continuous rain in Australia reflects the history of flooding in Bangladesh. The rain itself is a material trace of the hyperobject of climate change that binds the people of Bangladesh together with Henry and his descendants in England and Australia, just as it binds audiences at a dramatic performance. But this intimacy and repetition are followed by gaps, dark holes, and points of differentiation when the consequences of earlier events and actions result in far-reaching destruction and unknown futures, both in the human dimension and the material world. The impact of these events on the characters' personal lives separates father and son and engenders a prevailing mood of misunderstanding, disillusionment, and mystery in the play.

The materiality of the theatrical space is also significant: images and objects on the stage are not personified in an anthropocentric way, but have their own agency and dramatic effects. In the play, the raindrops and their sounds, umbrellas, raincoats, old suitcases, postcards, and fish are no longer in the service of humans, but each through its material representation is a "phasing" object or an "indexical sign" of the hyperobject of climate change in which it exists. These fleeting sensory perceptions do not formulate direct recognition or assure full knowledge of climate change, but each acts as a "*metonymy* for the hyperobject" (Morton, *Hyperobjects* 77). For example, as in Sewell's play, sound – in this case, the sound of the rain – plays a key role in dramatising "expanded materialism" via the play's form. The reality of climate change – matter at the most profound scale – cannot be perceived directly or captured by intersubjective or anthropocentric thinking. It exists in an "interobjective" mesh and the shared space of umbrellas hanging on the wall, the raincoats, the unstoppable rain, the characters, and the changing climate (see fig. 3.5). Thus, we can only discern

Figure 3.5 Umbrellas and raincoats hanging on the wall. *When the Rain Stops Falling*. Courtesy Black Swan State Theatre Company of WA, Perth, 2011. Directed by Andrew Mitchell. Photo credit: Gary Marsh.

some parts and pieces of it at a particular moment, or see its indexical signs. In this sense, the indexical signs are the hyperobject's hypocrisies acting in an ironic way because they do not present their own genuine existence. They pretend to be another thing while revealing "the hidden doom" of climate change, to use Morton's term (148). Thus, the time of hyperobjects involves a state of "hypocrisy": "With . . . their phasing in and out of human time and space, their massive distribution, their viscosity, the way they include thousands of other beings, hyperobjects vividly demonstrate how things do not coincide with their appearance" (Morton, *Hyperobjects* 174). I argue that these indexical signs in the play's theatrical space and their enmeshed relationship with the characters and with climate change animate the idea that things, humans, and nonhumans are not distinct entities but only exist through intra-action with each other. This perspective displaces human subjects from the centre of the play and allows nonhuman forces to surface through the traces that they are likely to leave in the lives of the characters.

In the 2008 Brink Production in Adelaide, the time-fractured narratives, inversion of logic, and coincidences are accompanied by the striking use of screens in the performance, so that the past, present, future, and different places are all staged simultaneously in a series of intertwining, multilayered scenes. The structural pattern of the play is broken up in such a way that the movement from one timeframe to another is not clear and a scene from one era enters into a scene from another. For example, when the stage set presents the red landscape of Uluru descending from the flies in 1988, Elizabeth, Henry's wife and Gabriel Law's mother, appears upstage as both a young woman in 1959 and an old woman in 1988 (see fig. 3.6). The scene reveals that the characters inhabit a spatiotemporal continuum. Also, layering multiple timeframes and scenes upon one another enables the audience to imagine the vastness of deep time onstage. Coincidences and recurrences cross generations as each person, oblivious to the others, enters and exits through a screen, hangs their raincoat and umbrella on a hook, looks out the window, sits down behind a single bench table, eats fish soup, and talks about the weather using the "figure of speech" about people drowning in Bangladesh. These concurrent patterns are staged in accordance with the innovative simplicity and flexibility of the design, which makes economic use of space by employing a variety of screens and some tables and chairs (see fig. 3.7). As the designer Valamanesh said, "[t]he complexity of the writing and its ever-changing settings required a space that could be flexible: from the intimacy of a domestic room to the grandeur of a vast natural environment" (Edwards and Kosky n. pag.). A remarkable feature of this production is that Valamanesh's design and Bovell's dramatic text were created collaboratively and so supplement each other (Gough-Brady). This interconnection between the text-based drama and the performance provides an opportunity to imagine ecological interconnection in a material way, since not only the actors but also the materials, objects, landscape, space, time, and weather play their own roles. In this way, I argue that the play stages

Figure 3.6 WTRSF_Rock_Wend Lear. Carmel Johnson as Elizabeth Law (older) and Michaela Cantwell as Elizabeth (younger) at the back of the stage. Yalin Ozucelik as Gabriel Law and Anna Lise Phillips as Gabrielle York (younger) at the front of the stage in the 2008 world premiere season by Brink Productions. Directed by Chris Drummond. Photo by Wend Lear.

agency in its own right, formulating a complex collaboration among the narration, the human characters, the climate, and the moment of performance. Challenging anthropocentric approaches to theatre, the play materially – through its immediacy, intimacy, and collectivity – creates an opportunity for thinking and talking about the large-scale, multifaceted event of climate change and so contributes to an awareness of our time, the Anthropocene. Further, the interconnection between text and performance acts as a formal analogy for the kind of critical and creative work the Anthropocene requires. It is this way of thinking about ecology and theatre that prompted the playwright and the designer to create a dramatic form based on questions like "How do we feel about the world? What do we think is going to happen in the world?" (Gough-Brady n. pag.). In other words, the play's collaborative genesis gives greater agency to the materiality of production, which in turn prompts important questions about human feeling and thinking. It is worthwhile to note that the play's 2016 production by artistic director Blanka Zizka at the Wilma Theater in Philadelphia deals with concern about the Anthropocene. Influenced by Roy Scranton's book *Learning to Die in the Anthropocene* (2015) and Rob Nixon's idea of "slow violence" the production revealed the play's implicit theme of the impact of slow violence on the family and on the climate (Bilderback).

Figure 3.7 *When the Rain Stops Falling* ©Jeff Busby_054. Each actor, oblivious to the others, enters and exits through a screen. From left, Yalin Ozucelik as Gabriel Law, Carmel Johnson as Elizabeth Law (older), Michaela Cantwell as Elizabeth Law (younger), Paul Blackwell as Joe Ryan, Anna Lise Phillips as Gabrielle York (younger), and Kris McQuade as Gabrielle York (older) in the 2009 return season at the Melbourne Festival by Brink Productions. Directed by Chris Drummond. Photo by Jeff Busby.

Meanwhile, the play evokes a sense of change, hope, love, the search for ecological understanding, care in the midst of despair, loss, dark secrets, the danger of extinction, torrential rain, and a massive flood in 2039. It does this through Andrew's quest for the truth about his father, "to know who he is, where he comes from, where he belongs" (Bovell, *Rain Stops Falling* 5), an act that gathers all four generations around the same table in the final moments of the play (see fig. 3.8):

> *The ancestors watch* ANDREW *as he wrestles with the weight of his mysterious past. And one by one, starting at the opposite end of the table, they place their hand on the hand of the ancestor who sits next to them until* GABRIEL LAW *places his hand on the hand of his grandson,* ANDREW. *And for this moment they are joined across time and continents.*
>
> (58)

Family members pass relics, including letters and postcards, from hand to hand, from ancestor to descendant, while looking at the extinct fish on the table – the

Figure 3.8 All the generations gather around the same table. *When the Rain Stops Falling*. Circa Theatre, Wellington, New Zealand, 2011. Directed by Susan Wilson. Photo by Stephen A'Court.

fish that Henry predicted would fall from the sky as a sign of the end of the world – and, more importantly, hearing the sound of the rain stop (see fig. 3.8). This inspiring moment of ecological awareness grows from the sense that the characters, like Donna Haraway's Children of Compost, "knew they could not deceive themselves that they could start from scratch. Precisely the opposite insight moved them; they asked and responded to the question of how to live in the ruins that were still inhabited, with ghosts and with the living too" (Haraway 138). This scene makes physically present for the audience the idea that remaining tied together will open up the possibilities of healing and ongoingness in a ruined environment.

The Earth has entered a new, melancholic age, many ecosystems have changed radically and many species are disappearing rapidly. These changes suggest that it is no longer possible to restore the environment to its former state. However, we need to at least understand the past and think about what we can change by dislocating dominant forms of anthropocentrism and abolishing our distance from the nonhuman world. The playwright, influenced by a Paris exhibition called *Melancholy: Madness and Genius in Art*, sees the play as melancholic not because it is about "warning and doom," but because melancholy is "the deep thinking we need to do that precedes change" (Bovell, *Rain Stops Falling* xiv). The play performs such a perception through a dramaturgical melancholic space of ecological interconnection that reduces temporal and spatial differences and raises the ecological consciousness of and shared responsibility for climate change. It is as if Bovell is able, through theatre, to counter the conventional state of ecological awareness, which is based on a normalising representation of environmental realities as self-regulating. In doing so, Bovell shows audiences how to live in the ruins we have made, and how understanding, change, love, promise, hope, peace, togetherness, and intimacy can unfold even in the dark, painful, and unsettling moments of life, with the capacity to move us on in a different way. While the end of the world is the beginning of ecological complexity and "hesitating futurality" (Morton, *Hyperobjects* 95), what is left behind is "intimacy": "We have lost the world but gained a soul – the entities that coexist with us obtrude on our awareness with greater and greater urgency" (108). Although the world has come to its end, the world of the play is enveloped in melancholy as the rain stops falling, leaving the audience to think about ecological interconnectedness.

In exploring the ways in which selected Australian plays grapple with the representational intrusions and displacements of global environmental change, this chapter has demonstrated how the hyperobject of climate change, which acts as an expanded agential force across a field of human bodies and nonhuman objects, displaces the conventional human-centred form of drama. Interrupting the flow of electrical power in Sewell's play and hearing the sound of the rain stop in Bovell's play are notable moments that show how a theatrical work can effectively and innovatively deepen its audience's ecological awareness, if only for a moment. These contemporary works of Australian theatre contribute to the theatre of climate change, prompting us to act decisively in the face of the growing climate emergency.

Notes

1 See Kate Rigby's "Weaving the environmental humanities: Australian strands, configurations, and provocations" (2019).

2 See Helen Gilbert and Jacqueline Lo's *Performance and Cosmopolitics: Cross-Cultural Transactions in Australasia* (2009) and Joanne Tompkins' *Unsettling Space: Contestations in Contemporary Australian Theatre* (2009).

3 In his conference paper "Materialism Expanded and Remixed," Timothy Morton argues that "[t]he creation of what I've called *hyperobjects* also causes us to rethink materialism. Hyperobjects are . . . developments in quantum theory [and] require us to rethink matter at the most profound scale we have yet imagined" (2).

4 Since the start of the Anthropocene epoch, Timothy Morton argues, art has entered a new phase of aesthetics, "the Asymmetric Phase," in which "our cognitive powers become self-defeating" (*Hyperobjects* 160). Although the name "Anthropocene" has been taken from humans due to their part in creating this age, the related knowledge of emergent outcomes are beyond and resistant to human comprehension, decentralising human agency.

5 According to Timothy Clark, the movie *The Day After Tomorrow* contains plots of "individual decision or heroism that reinforce a culture of narcissistic individualism" ("Some Climate Change Ironies" 144).

6 Timothy Morton argues: "You can no longer have a routine conversation about the weather with a stranger. The presence of global warming looms into the conversation like a shadow, introducing strange gaps. . . . A hyperobject has ruined the weather conversation, which functions as part of a neutral screen that enables us to have a human drama in the foreground" (*Hyperobjects* 99).
7 The play is also an appropriate description of Rob Nixon's question: "How do we bring home – and bring emotionally to life – threats that take time to wreak their havoc, threats that never materialize in one spectacular, explosive, cinematic scene?" (*Slow Violence* 14).
8 According to David Lindenmayer, the Murray-Darling Basin constitutes about one-seventh of the whole Australian continent and is considered one of the most important areas for agricultural and grazing activities (64). However, excessive land clearing in the area has caused a salinity problem and there is some evidence to suggest that "by 2020 an estimated 630,000 hectares (2.4 per cent) of the basin will have water tables within 2 metres of the surface" (64). The risk of irreparable damage is substantial as "[t]he salinity of these waterlogged areas will be chronic and the land will be ruined, both for agricultural production and for many species of animals and plants" (64).

References

Bilderback, Walter. "Fish Soup, Mourning, and Hope at the End of the World." Web blog post. *HowlRound*. HowlRound, 18 Apr. 2017. Web. 23 June 2017. <https://howlround.com/fish-soup-mourning-and-hope-end-world>.

Bovell, Andrew. "The Zeitgeist of the 21st Century." Web blog post. Interview by Walter Bilderback. *Wilma*. Wilma Theater, 11 Oct. 2016. Web. 18 Dec. 2016. <www.wilmatheater.org/blog/zeitgeist-21st-century>.

———. *When the Rain Stops Falling*. Sydney: Currency P, 2009. Print.

Bramwell, Murray. "Weathering the Past and Reconciling the Future: Andrew Bovell's *When the Rain Stops Falling*." *When the Rain Stops Falling*. Ed. Andrew Bovell. Sydney: Currency P, 2009.xiii–xvii. Print.

Brecht, Bertolt and John Willett. *Brecht on Theatre: The Development of an Aesthetic*. New York: Hill and Wang, 1964. Print.

Chakrabarty, Dipesh. "The Climate of History: Four Theses." *Critical Inquiry* 35.2 (Winter 2009): 197–222. Print.

Chaudhuri, Una. "Dramaturg's Notes: The Weather Changes." *Research Theatre, Climate Change, and the Ecocide Project*. Eds. Una Chaudhuri and Shonni Enelow. New York: Palgrave Macmillan, 2014.121–3. Print.

Clark, Timothy. "Some Climate Change Ironies: Deconstruction, Environmental Politics and the Closure of Ecocriticism." *The Oxford Literary Review* 32.1 (2010): 131–49. Print.

Colquhoun, Lachlan. "Drama of Climate Change." *The Adelaide Fix* 29 Feb. 2008: 4. *Rip It Up*. Web. 10 May 2014. <http://theadelaidefix.com.au/images/the_fix_004.pdf>.

Crowley, Kate. "Is Australia Faking it? Kyoto Protocol & the Greenhouse Policy Challenge." *Global Environmental Politics* 7.4 (2007): 118–39. Print.

Drummond, Chris. "Finding the Life in the Stone." *When the Rain Stops Falling*. Ed. Andrew Bovell. Sydney: Currency P, 2009.vii–xii. Print.

Edwards, Robyn and Samantha Kosky. "Teacher's Resource Kit." *Sydney Theatre Company's When the Rain Stops Falling Notes*. Sydney Theatre Company, 2009. May 2014. Web. 15 May 2015. <https://australianplays.org/assets/files/resource/doc/Ruby_Moon_Teachers_Kit.pdf>.

Elliott, Peter, Ned Manning, Margaret Saltau and Elizabeth Surbey. *Drama Reloaded*. Melbourne: Cambridge UP, 2011. Print.

Flannery, Tim F. *The Future Eaters: An Ecological History of the Australasian Lands and People*. Chatswood, NSW: Reed Books, 1994. Print.

Gilbert, Helen and Jacqueline Lo. *Performance and Cosmopolitics: Cross-Cultural Transactions in Australasia*. London: Palgrave Macmillan, 2009. Print.

Glow, Hilary. "Looking into the Abyss." *It just Stopped: Myth, Propaganda and Disaster in Nazi Germany and Contemporary America*. Ed. Stephen Sewell. Sydney: Currency P, 2007. vii–xiv. Print.

Gough-Brady, Catherine. "Extreme Weather Alert." *Newmatilda.com*. New Matilda, 28 Oct. 2009. Web. 14 May 2014. <https://newmatilda.com/2009/10/28/extremeweather-alert>.

Haraway, Donna. *Staying with the Trouble. Making Kin in the Chthulucene*. Durham, NC: Duke UP, 2016. Print.

Lester, Rebecca E. and Peter G. Fairweather. "Ecosystem Response Modelling of the Murray Mouth and Coorong Lagoons." *Ecosystem Modelling in the Murray-Darling Basin*. Eds. Neil Saintilan and Ian Overton. Collingwood: Csiro Publishing, 2010.373–90. Print.

Lindenmayer, David. *On Borrowed Time: Australia's Environmental Crisis and What We Must Do about it*. Camberwell: Penguin Books, 2007. Print.

Malloch, Lachlan. "The Nightmare Is Now." *Green Left Weekly*. Democratic Socialist Party, 14 Oct. 2006. Web. 16 Aug. 2014. <www.greenleft.org.au/node/36434>.

Morton, Timothy. "Art in the Age of Asymmetry: Hegel, Objects, Aesthetics." *Evental Aesthetics* 1.1 (2012): 121–42. Print.

———. "Dark Ecology: Philosophy in the Anthropocene." Humanities Without Boundaries. UW-Madison Center for the Humanities, Madison. 19 Oct 2012. Lecture.

———. *Ecology without Nature: Rethinking Environmental Aesthetics*. Cambridge: Harvard UP, 2007. Print.

———. *Hyperobjects: Philosophy and Ecology after the End of the World*. Minneapolis: University of Minnesota P, 2013. Print.

———. "Materialism Expanded and Remixed." *New Materialisms*, Johns Hopkins University, Apr. 13–14 2010: (1–17). Scribd. 04 July 2015. <http://dm.ncl.ac.uk/courseblog/files/2011/03/25830212-Materialism-Expanded-and-Remixed.pdf>.

———. "OOO and Quantum Theory." Web blog post. *Ecology Without Nature*. Timothy Morton, 27 July 2013. Web. 02 Nov. 2016. <http://ecologywithoutnature.blogspot.com.au/2013/07/ooo-and-quantum- theory.html>.

———. *The Ecological Thought*. Cambridge: Harvard UP, 2010. Print.

———. "Zero Landscapes in the Time of Hyperobjects." *Graz Architectural* 7 (2011): 78–87. Web. 12 Feb. 2014.

Mostafa, Meraz and Pablo S. Bose. "The Narrow Focus on Climate Change in Bangladesh often Reproduces Exploitation and Vulnerability rather than Addressing it." Web blog post. *South Asia @ LSE*. London School of Economics, 24 Oct. 2016. Web. 15 Dec. 2016. <http://blogs.lse.ac.uk/southasia/2016/10/24/the-narrow-focus-on-climate-change-in-bangladesh-often-reproduces-exploitation-and-vul-nerability-rather-than- addressing-it/>.

Nixon, Rob. *Slow Violence and the Environmentalism of the Poor*. London: Harvard UP, 2011. Print.

Payne, Charlie. et al. "Almeida Projects: *When the Rain Stops Falling* by Andrew Bovell." *Yumpu*. Yumpu.com, May 2009. Web. 20 Feb. 2019. <www.yumpu.com/en/document/read/46425228/when-the-rain-stops-falling-almeida-theatre>.

Rigby, Kate. "Imagining Catastrophe: Utopia and Dystopia in a Warming World." *Changing the Climate: Utopia, Dystopia and Catastrophe*. Eds. Andrew Milner, Simon Sellars and Verity Burgmann. Melbourne: Arena Publications, 2011.57–77. Print.

———. "Weaving the Environmental Humanities: Australian Strands, Configurations, and Provocations." *Green Letters* 23.1 (2019): 5–18. Print.

Robin, Libby. *How a Continent Created a Nation*. Sydney: University of New South Wales P, 2007. Print.

Seton, Mark. *It Just Stopped and Myth, Propaganda and Disaster in Nazi Germany and Contemporary America: A Drama in 30 Scenes*, by Stephen Sewell. Book review. *Australasian Drama Studies* 66 (Oct. 2008): 230–2. Print.

Sewell, Stephen. *It Just Stopped/Myth, Propaganda and Disaster in Nazi Germany and Contemporary America*. Sydney: Currency P, 2007. Print.

———. "So is it the End of the World, or What? Stephen Sewell on his Apocalyptic Comedy It Just Stopped." Interview by Run-Riot. *Run Riot*. 3 Jan. 2014. Web. 01 May 2014. <www.run-riot.com/articles/blogs/so-it-end-world-or-what-stephen-sewell-his- apocalyptic-comedy-it-just-stopped>.

Stevens, Lara, Peta Tait, and Denise Varney, eds. *Feminist Ecologies: Changing Environments in the Anthropocene*. Switzerland: Palgrave Macmillan, 2018. Print.

Tait, Peta. *Fighting Nature: Travelling Menageries, Animal Acts and War Shows*. Sydney: Sydney UP, 2016. Print.

———. *Wild and Dangerous Performances: Animals, Emotions, Circus*. London: Palgrave Macmillan, 2012. Print.

Tompkins, Joanne. *Unsettling Space: Contestations in Contemporary Australian Theatre*. London: Palgrave, 2006. Print.

Usher, Robin. "Two People, Backs to the Wall." *The Age*. The Age Company, 04 Apr. 2006. Web. 20 June 2015.

Varney, Denise. "'Beauty Tigress Queen': Staging the Thylacine in a Theatre of Species." *Journal of the Association for the Study of Australian Literature* 15.2 (2015): 1–13. Print.

———. *Radical Visions 1968–2008: The Impact of the Sixties on Australian Drama*. Amsterdam: Rodopi, 2011. Print.

4 Setting the stage for material expressions across Planetary boundaries

E.M. Lewis's *Song of Extinction* and Chantal Bilodeau's *Sila*

The previous chapter argued that theatre based on themes of family, nation, and politics can be productively reread from an ecocritical perspective to demonstrate theatre's potential for challenging the presumption that humans are inherently separable from nonhumans. However, the plays discussed in this chapter, including *Song of Extinction* (2010) by the American playwright E.M. Lewis and *Sila* (2015) by the Canadian playwright Chantal Bilodeau, have been chosen for their critical and ecological consciousness of contemporary concerns about global environmental risk. In North America, environmental humanities has emerged as a fast-expanding field characterised by a variety of themes and topics relevant to specific regions, including "climate change and glacial melt in Canada, oceans on the West Coast, deserts in the Southwest, legacies of postindustrial decline in the Midwest, urban environment in large cities, and relations with Indigenous/First Nation groups across Canada and the US" (O'gorman et al. 434). These themes are mainly situated in environmental justice modes of inquiry by involving the experiences of people of different cultures, values, histories, races, genders, and classes. The study of these environmental concerns in plays and performances, however, has principally been on the periphery of mainstream theatre in North America, particularly on larger stages.

The early 1990s saw the first publications on ecocritical theatre in the United States as theatre started embracing growing concerns over environmental crisis. It is only in the last few years that the number of climate change plays in the United States has begun to increase. In this respect, Bilodeau refers to a HowlRound article (2019) by Marshall Botvinick as an evidence that shows how much the US theaters have been reluctant to engage with climate change. As Bilodeau writes, the article "looked at U.S. world premieres in the 2019–20 season at 75 member theatres of the League of Resident Theatre and the 32 National New Play Network core member theaters [while] exactly zero percentage of plays had climate change or the environment as a thematic focus" ("Digging for New Roots" n. pag.).[1] Despite this "depressing" statistic, Bilodeau refers to the emergence of a few exciting new works by American theatremakers and companies, including PearlDamour's *Ocean Filibuster* (not yet premiered), Phantom Limb Company's trilogy of works – *69°S* (2011),

DOI: 10.4324/9781003048749-4

Memory Rings (2016), and *Falling Out* (2018), Superhero Clubhouse's *Mammelephant* (not yet premiered), and Siren Arts's *Mermaid Truth or Consequences: A Biodiversity Game Show* (2018).

In regard to Canada, Ella Soper and Nicholas Bradley's anthology, *Greening the Maple: Canadian Ecocriticism* (2013), suggests that Canadian ecocriticism has emerged out of a uniquely Canadian understanding of literature and the environment between the 1960s and 1980s, on the one hand, and American ecocriticism in the 1990s on the other. Despite the solidification of ecocriticism in Canada, in a special issue of *Canadian Theatre Review* (2010) that was devoted to theatre and ecology, Nelson Gray and Sheila Rabillard write that "[e]cologically informed discourse about theatre and performance has, with only a few exceptions, been strangely absent in Canada" (3). The writers state that the conjunction of regional and national sources and performance is a rich site of inquiry for Canadian ecocriticism. They also proceed to "act globally" by focusing on plays and performance projects that link Canada to the planet. The 2020 issue of *Canadian Theatre Review* that deals with "Extractivism and Performance" represents an expansion of the theatrical scope of Canadian ecocriticism. In the editorial section of the issue, Kimberly Richards and Heather Davis-Fisch argue that many Canadian plays and performance practices "have dramatized how the natural resources of the land, and the [extractive] industries developed around them, have shaped social values, labour practices, political alliances, and notions of citizenship" (5). The writers cite a few well-known works, including Ted Johns and Paul Thompson's *The Farm Show* (1998) which depicts agricultural practices in Clinton, Ontario, Marie Clements's *Burning Vision* (2003) which represents the radiating effects of uranium mining, Annabel Soutar's *The Watershed* (2016) which investigates into the Harper administration's environmental policies, along with *The Pipeline Project* (2020) by Kevin Loring, Quelemia Sparrow, and Sebastien Archibald, dealing with the tensions between Indigenous communities and the descendants of settlers through the environmental debate on pipeline expansion on unceded territory (6).

In this chapter, I focus on two North American playwrights whose works were written and are performed with the intention of representing ecological issues. *Song of Extinction* and *Sila* engage explicitly with the future of the earth and all life, not just that of the human species. They each won the Earth Matters on Stage Ecodrama Award, in 2009 and 2012 respectively, under the label "ecodrama." As Wendy Arons and Theresa May contend, these plays have "ecodramaturgical" features "that reside . . . in the work itself and that [are] (also) a critical strategy" because they foreground and illuminate ecological issues and themes, dramatising the reciprocal relationship between humans and the more-than-human world (181). On this point, for the playwright Bilodeau, "[a] new consciousness requires new artistic constructs" that "embody, in form, content, and process, the essence of the issues we are facing" ("New Aesthetic" n. pag.). Given the necessity of different theatrical forms in the Anthropocene epoch, these plays provide appropriate and responsive instances of Una Chaudhuri's "fifth wall dramaturgy," which insists on dismantling the fifth wall of anthropocentrism and activating "the 'diegetic' mode of literary communication" rather than "the 'showing,' or the 'mimetic' mode" ("Fifth Wall" n. pag.). I argue that this innovative and experimental mode of communication opens up an understanding of the possibilities and strategies of ecocritical theatre – strategies made necessary by mass extinction, climate change, and the critical issues of agency and scale that they entail. In order to see how *Song of Extinction* and *Sila* contribute to the development of the kind of ecodramaturgy necessary for thinking about and narrating the Anthropocene, I focus on how they tell stories and use imagery that prompts a deeper exploration of our own ecological identity and condition, not only beyond human-nonhuman divisions but also across agential, spatial, temporal, and geographical boundaries.

1 Dramatising the end of the species in the deterritorialized space of *Song of Extinction*

Song of Extinction premiered at the Moving Arts Theatre Company in Los Angeles in November 2008. The narrative follows a series of catastrophic events in human and natural environments comprising a classroom and a hospital in contemporary Portland, Oregon; the Cambodian killing fields in 1970; and the Bolivian rainforest. It weaves together the stories of Max Forrestal, a troubled and isolated high school student and viola protégé; his biology teacher, Khim Phan, who is a Cambodian refugee; Max's father, Ellery Forrestal, a field biologist; and his mother, Lily, who writes high school science textbooks and is dying of cancer. While Ellery is intent on – and, to his son, seems obsessed with – conserving a species of rare, endangered Bolivian insect and its forest habitat, Max enters a state of deep sadness and emotional shock about his mother's approaching death. While attempting to reach out to his troubled student, Khim is haunted by his own past. Flashbacks and first-person narration reveal that Khim is the sole member of his family who survived the Khmer Rouge killing fields of the 1970s. Khim's loss of his family to war at the age of fifteen is contrasted with Max's alienation from his own family at the same age, as the play provides both personal and global perspectives on trauma. As Lily's condition deteriorates, Max runs away, and Khim looks for her at the hospital. There, in a dream sequence, he helps Lily return to Bolivia, to a time and place where her life was filled with love and hope. Influenced by Khim's story of the mass executions on the killing fields, Max writes a biology paper that he titles "Song of Extinction," though it is not simply a song about "becoming extinct" but also a song about "becoming" (Lewis, *Song of Extinction* 84).

The play's narrative is founded upon an innovative approach to extinction that revolves around different stories woven together into an interconnected pattern. The play broadens the sense of extinction by asking how thinking through stories of extinction – human and nonhuman – across spatiotemporal boundaries provides a microcosmic prism for the large-scale and unimaginable phenomenon of the sixth mass extinction. These ecodramaturgical inquiries resonate within the play in tandem with the three senses of extinction discussed by Claire Colebrook:

> the now widely discussed sixth great extinction event (which we have begun to imagine and witness, even if in anticipation); extinction by humans of other species (with the endangered species of the 'red list' evidencing our destructive power); and self-extinction, or the capacity for us to destroy what makes us human.
>
> (*Death of the PostHuman* 9)

The play enacts these different but interconnected scenarios that play out on personal, local, national, and global scales. Khim's lesson is about the sixth extinction, while Ellery's insect is an example of the second kind where humans directly destroy a species that we can identify, and the Cambodian genocide is an example of the third sense of extinction. Of course, what they all share is the belief that the pervasive impact that humans have on the fate of other species and the Earth's geological and biological history might lead to humankind's own extinction.

Extinction first appears when Khim is teaching the topic. Khim, played by an Asian-American actor in the 2011 performance in Theater Latté Da in Minneapolis, tells us that this is the most difficult part of biology for him to teach:

> There are things I know about extinction I don't know how to tell to my students. . . . Most chapters are easy for me to teach. DNA – easy. Cell structure – easy. This – I think too much. I tell myself, in my head, "It is not about you, dummy." (*beat*) Extinction is a very messy business. In books, it looks clean. First there is this animal. Then this animal goes away. But I remember extinction and it was not clean.
>
> (8)

Different definitions of extinction are embedded in Khim's language here. On the one hand, extinction refers to the disappearance of a nonhuman species. On the other hand, Khim refers to his experience of the Cambodian genocide and racial discrimination, as he explains in later scenes. Khim's statement expands the sense of extinction, implying that it is not a singular process. As the editors of *Extinction Studies: Stories of Time, Death, and Generations* (2017) assert, "there is no singular phenomenon of extinction; rather, extinction is experienced, resisted, measured, enunciated, performed, and narrated in a variety of ways to which we must attend" (Rose et al., "Telling Extinction Stories" 2–3). Extinction studies, therefore, focuses on "the 'entangled significance' of extinction" and "understanding and responding to processes of collective death, where not just individual organisms, but entire ways and forms of life, are at stake" (Rose et al., "Telling Extinction Stories" 3, 5). I argue that these concerns inform the play's extinction stories, insisting that the stories are inherently and inextricably "attentive to the simultaneously biological and cultural complexity of our world" (Rose et al., "Telling Extinction Stories" 5). The apparently divergent senses of extinction are represented as entangled with each other throughout the play, and so make extinction complicated, messy, and difficult to understand. For example, the usual meaning of extinction is exemplified by the endangered Bolivian insect that is part of Ellery's biology experiment and whose image is projected on the kitchen wall of the Forrestal home. This scene about the insect is followed by one in which Lily, fragile and lying in a hospital bed, is told that she will die in less than a week's time. Lily's struggle with death takes place as Ellery enters into a desperate struggle with the businessman Gill Morris to stop his company from destroying the Bolivian rainforest and while Khim relives the genocide in his native Cambodia. The play engages these diverse and interwoven stories, deals with the urgency of approaching extinction as both biological and cultural, and so reflects on diverse people's responses to and relationships with the changing world.

We can refer to these juxtapositions as the play's ecodramaturgical structure. Through its form, the play argues that humans are no longer defined by territorial or geographical criteria but have become planetary citizens with responsibility for conserving the Earth. In the 2008 performance at the Moving Arts Theatre Company in Los Angeles, set designer Stephanie Kerley Schwartz ingeniously interpreted the playwright's words in design elements including column-high shrines and an overhead screen, which allowed different images to represent location changes (Citron). This created a "deterritorialized" sense of place, to use the term from Gilles Deleuze and Félix Guattari that Ursula K. Heise reworks, moving the audience's gaze from Bolivia to Cambodia and then to a hospital and classroom in Portland. In another example, for the play's 2011 production by Theatre Latté Da at the Guthrie Theatre in Minneapolis, set designer Michael Hoover made the play's present time the focus of its revised idea of extinction by placing a hospital bed at the centre of the work. It was surrounded by transparent walls with panels inside them for representing other stories of extinction at other local and global scales. These productions provide new aesthetic possibilities for imaging a "deterritorialized environmental vision" through which, as Heise argues, we can see the interdependence between the global and the local and, more specifically, the interaction among cultural and ecological systems that are reconfigured at different scales and within larger global networks (*Sense of Place* 10, 51).

The play's demonstration of local and personal events of loss parallels a broader framework of extinction. Amid the stories of Bolivia, Cambodia, and Portland, the audience sees Khim in front of the board in his empty classroom, practicing his lesson on extinction in another sense – the mass extinction crisis:

> During the whole life of this planet, between five and fifty billion species have existed. T. L. Erwin says there are maybe . . . forty million different species of animals and plants on earth right now. Normal extinction level,

> background level, is one species gone every four years. Right now? Thirty thousand species disappear every year. Become extinct.
>
> (Lewis, *Song of Extinction* 33)

Khim's lesson positions the play within the current discourse on extinction: that the Earth is entering a new epoch that biologists and scientific communities label as the sixth mass extinction. For example, in her book *The Sixth Extinction: An Unnatural History* (2014), Elizabeth Kolbert writes that by drastically and profoundly altering the shape and future of life on the planet, humans have triggered a sixth mass extinction. She reconsiders five prehistoric mass extinction events spanning five hundred million years and compares them to the rapid, widespread, and elevated extinction rates at the turn of the third millennium. The sixth extinction is predicted to be mankind's most devastating and lasting legacy, carried on by growing threats of habitat destruction, exploitation, climate change, and more. So, rather than being a natural process, this large-scale biological transformation induced by human activities is a manifestation of the Anthropocene. This key point, which is embedded in Khim's lesson, has a significant function in developing the play's illustration of the broad concept of extinction. Keeping this point in mind, I would like to place the play in a group of stories that, as Thom van Dooren argues, we need for the Anthropocene. According to van Dooren, these are "the stories that take the complexity of *change*, and draw in some of the myriad beings – human and not – for whom this change is all too often experienced as suffering and loss" ("Vulture Stories" 55). Such narrations also "ask their audiences to be curious and to care about the many relationships, the many ways of being, [and] the many worlds that are disappearing in bright bursts of pain in this time of escalating extinctions" (55). In the following sections, I consider how the play's web of intricately overlapping losses presents the audience with a range of temporal and spatial scales in which the Cambodian killing fields, Bolivian deforestation, and the death of a mother in Portland are juxtaposed with the dark global picture of mass extinction that Khim describes.

Bolivia's ecocide: the critique of modernity

The relationship between the risk of biodiversity loss and the processes of modernisation and industrialisation is evident in a conversation between Ellery, the biologist, and Gill, the developer. In a lecture on the endangered state of an endemic insect subspecies, Ellery shows a slide of the Dynastidae Strategus of the Banados del Izozog region of Bolivia, which he has "discovered" and has been researching for twelve years. However, putting insects at risk and causing their extinction is an inevitable consequence of Gill's pursuit of economic profits. Such profit-seeking motivates Gill to develop the Bolivian rainforest, which Lily describes as "[m]iles from civilization" (Lewis, *Song of Extinction* 48), and make the forest useful for industrialisation, global marketing, and capitalism. Gill justifies deforestation and harm to rainforest species with the claim that the loss of species is replaced with "the infusion of 7.9 million dollars into the Bolivian economy; not to mention what it will do for ours. Lumber. Houses" (16). Job security and the relentless pursuit of economic growth are put before long-term environmental protection and ecological sustainability. The choice of Bolivia as the representative site of deforestation is significant given its known ecological vulnerability. The region's environmental degradation and disappearance of species cannot be counted as natural; rather, they are the result of deforestation and human disturbance. Bolivia is regarded as one of the "mega-diverse" countries, and its rich environment with wide tracts of tropical forest and high biodiversity has been critical in developing the country (Slunge and von Walter 11). However, as Daniel Slunge and Susanne von Walter assert, "the rapid industrial development in mining, gas, oil and agriculture experienced in Bolivia . . . is associated with very high environmental costs in terms of pollution, deforestation and land degradation" (4). Correspondingly, the play's representation of the endangerment of the Bolivian insect expresses "unease with modernization processes" or reveals "an explicit critique of modernity and the changes it has brought about in humans' relation to nature" (Heise, *Imagining Extinction* 32). Human-driven biodiversity loss is among the uncompensated environmental costs that make the whole region, including its humans and nonhumans, vulnerable to ecological change.

For Ellery, the horror of Bolivia's ecocide and the extinction of an entire species of beetle affect human-nonhuman interaction and might go hand-in-hand with further ripples of degradation and suffering:

ELLERY: Wilson – from Harvard, E.O. Wilson – predicted that half of all species will be extinct in 50 years if current land use patterns continue.
GILL: It's a bug, Mr. Forrestal.
ELLERY: It's genocide.
GILL: You are being absurd!
ELLERY: You are killing an entire species.
GILL: The word "genocide" refers to people. . . . It is a bug. If it was walking across my desk right now, I'd squish it with my *New York Times*. I don't care! You want to put two hundred and fifty Bolivian workers out of a job and let them starve to death because of a bug?
ELLERY: It's not about them. . . . The potential implications of the extinction of a species are . . . devastating. You can't know . . . what we might be losing if we were to lose an entire species. The cure for . . . The cure for cancer might be – (Lewis, *Song of Extinction* 15)

The businessman's ephemeral economic benefits confront the biologist's recognition of deep-rooted environmental costs and unprecedented destruction of life. Tracing the "potential implications" that may arise from the neoliberal exploitation of the environment is very complicated because both humans and nonhumans are affected across space and time. For example, regarding the recent Bolivian forest fires and its close relationship to the political economy of extraction, Manuela Picq argues that ecocide inevitably entails genocide:

> The government's claims that there are not 'enough deaths' is a horrendous dehumanization of indigenous lives. It is also against the rights of nature, guaranteed in the 2009 Constitution, and the notion that humans are part of nature. . . . Indigenous [Bolivian] peoples live in symbiosis with their ecosystem. . . . To burn forests that are home to dozens of endemic species is not only to destroy basic resources but to kill members of Indigenous communities, even if they are not human. For [Bolivians], these man-made forest fires are literally genocide because they destroy their cultures, economies, language and land. In times of climate crisis, ecocides have genocidal effects because they destroy vital relationships especially between humans and nature.
>
> (n. pag.)

Extractive land policies may even affect humans in other parts of the world. Deforestation, for instance, has a close connection with Lily's death in a different country, not only in terms of the disappearance of species that could be a primary source of medical treatment for her disease but also because of Lily's emotional bond with Bolivia's natural landscape. An example that illustrates this connection is Ellery's dream in which he sees their idyllic past in Bolivia, where Lily wrote her scientific textbook and he watched "the Dynastidae Strategus hatch from their tiny clutches of eggs" (Lewis, *Song of Extinction* 58). In his dream, he asks Lily to go back to their tent by the time the trees at the far edge of the jungle begin to fall – a sign of Gill's deforestation. He is frightened and struggles to protect Lily from harm: "ELLERY *looks wildly toward the place where the chopping sound is coming from, then picks up a piece of firewood. He puts himself between the chopping sound and the tent*" (50). The parallel between this dream and Lily's endangered present life reveals that the anthropogenic degradation of the forest affects humans in a complicated way (see fig. 4.1). It it worth noting that this is not to say that the extinction of a nonhuman species, genocide or individual death are identical, but rather, in sync with Deborah Bird Rose, "to pose connections that help us understand both more deeply" ("Monk Seals" 136). For Rose, "the anthropogenic mass-extinction event now in process is to be understood as another human-caused mass-death event. . . . [I]t is possible and ethically significant 'to think human and animal deaths as capable of illuminating each other in their separate darknesses'" (136), and, I add, in the face of broader ecological transformations that the Anthropocene sums up and in which humans are entangled with the nonhuman world.

Lewis's work also shows how the story of an endangered species needs to be considered biologically and culturally as it plays into how we understand modernity and shapes the science of extinction. The play's dramatisation of extinction in Bolivia, through its focus on the tension between industrial capitalism and ecological issues, evokes what Heise calls "turning points in human

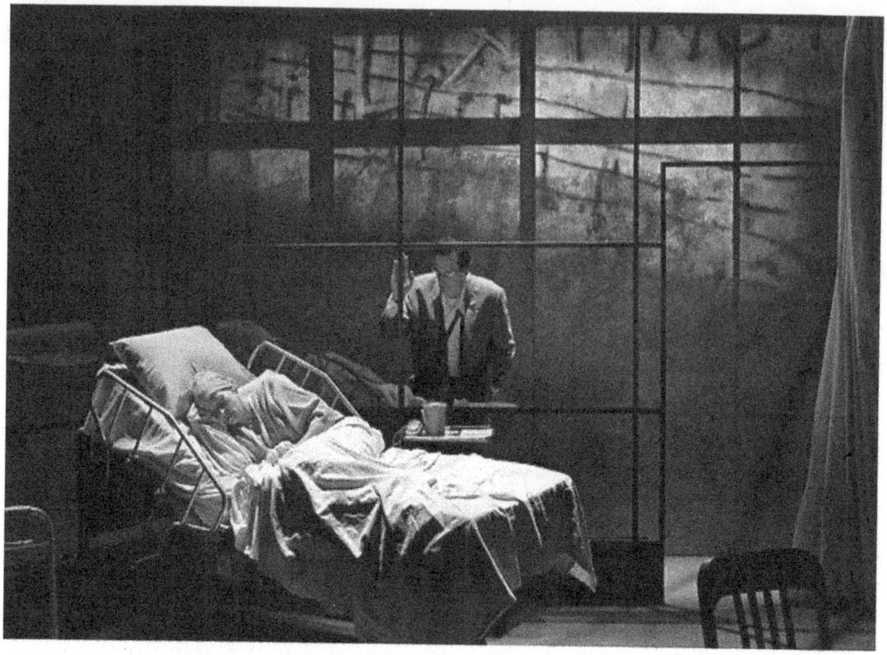

Figure 4.1 John Middleton as Ellery and Carla Noack as Lily. *Song of Extinction*. Theatre Latte Da, Minneapolis, 2011. Directed by Peter Rothstein. Photo credit: Michal Daniel.

cultural histories, in which the loss of a particular species stands in both for a broader sense of the vanishing of nature and the weakening of human bonds to the natural world," as well as "hopes that a part of one's own identity and culture might be preserved or revived if an endangered species could be allowed to survive or an extinct one be rediscovered" ("Lost Dogs" 69). Along with the disappearance of the insects, part of human identity and culture itself seems to be lost – not only Lily's health but also Ellery's identity as a biologist.

Cambodia: ecocide in the shadow of genocide

Alongside its representation of the Bolivian ecocide, the play reflects on the Cambodian genocide of the twentieth century. It examines this genocidal act through tenuous conversations between Khim and Max, the latter in a state of self-imposed isolation and stupor due to the approaching death of his mother and the failure of his relationship with his father. On the one hand, Max resists making a conscious connection to the subject of extinction despite the word *EXTINCTION* being written in large red letters on the whiteboard in his classroom. Khim, on the other hand, claims, "I know something about extinction" (Lewis, *Song of Extinction* 52), referring to his first-hand experience of the genocidal killing fields of Cambodia. In his American high school classroom, he now teaches consideration and respect for endangered nonhuman species while also implicitly invoking Cambodia: "I do not talk about it ever. But in this chapter on extinction, my country and history and family come up bitterly in my throat. And maybe I teach them about Cambodia without ever saying its name" (47). In a broader perspective, his concern for nonhuman species at risk becomes for him a projection of an integrated self – an organism who embodies biology, family, community, history, and culture in a way that enfolds human genocide in the wider framework of species extinction.

Yet Cambodia's tragedy was not limited to crimes against humanity. It included environmental genocide that, historically, has not received attention. Cambodia's political and ecological trajectory during the rule of the Khmer Rouge points to the possibility that genocide and ecocide are connected. The play marks the anthropogenic disturbance of land, natural resources, and species as modernity advances into the third world. The ecocide in Bolivia, as described by Ellery, provides the key example of this process. Anthropogenic activities have affected Cambodian timberlands, which have served a variety of ecological, economic, social, and political purposes in both local and global contexts. In his essay, "(Eco)Genocide in Cambodia," Forrest McGraw writes that the timberlands, which were once "a provider of sustainable life," faced increasing degradation after the hostile reign of the Khmer Rouge in 1975 (n. pag.). McGraw also considers the political dimensions of the commodification of Cambodian forests and how the global tropical timber trade financed the military activity of both the Khmer Rouge and the Royal Government of Cambodia. Such international trade affected not only "forest-based resources" and ecological systems but also induced biodiversity loss and endemic extinctions alongside the better-known acts of genocide (McGraw n. pag.). Although the play does not directly make this connection, Khim's memory of his lost homeland – where he "stood, once, in a field beneath the Elephant Mountains, wide green field that smelled like grass under a wide blue sky and butterflies with yellow wings, glimmer glimmer in the sunlight came down and lit on [him]" (Lewis, *Song of Extinction* 8) – might indicate the way genocide is entangled with ecocide. Khim's narration of the ruined harmony of the land reveals that genocide should be regarded as a preface to unprecedented environmental degradation.

Through reference to his tragic life history and the Cambodian (eco)genocide, Khim makes an intimate connection with extinction. Human self-extinction as well as the extinction of other species cannot be merely confined to abstract concepts; rather, in order to be understood, they need to be defined as personal and cultural concerns. This connection reminds me of Rose's description of her book *Wild Dog Dreaming: Love and Extinction* (2011) in her 2018 interview with van Dooren. In writing her book, Rose draws upon insights from Old Tim Yilngayarri – an Indigenous clever man and one of Rose's most generous teachers – and his kinship with animals. The other inspiration for Rose is the endangered wild dogs of Australia known as dingoes that have an intricate relationship with Aboriginal people. Dingo's story is disturbing because the threat to their future is being actively planned by humans.[2] Rose says:

> 'It was talking to those dingo dreaming guys, [talking] with Old Tim, that I realised, he has been through slaughter and he's a survivor of genocide. He's a survivor in a really particular and peculiar way, and he is sitting here witnessing the war against dingoes. He is seeing another round of extinctions. He knows the white guys would just like to get rid of dingoes forever more, so he is seeing it happen now to his nonhuman kin. That, I think, was what drove me to the understanding that you can't sit and listen to this stuff in an innocent way, you can't be detached from it, you can't be over involved with it either (it's not your story); and so you have to listen in a really different, attentive, way.'
>
> (van Dooren, "A Howling Lamentation" n. pag.)

This description of the intertwined process of genocide and ecocide in the face of the ongoing mass extinction of our times calls for being attentive to the changing world – through our own or others' stories – and its possibilities for living and dying. For Khim, in the dark scene of Cambodia's tragedy, when people disappeared in the most obvious way, "there can be no more pretending" (Lewis, *Song of Extinction* 62). Similarly, he says that Max is currently in "the place of no more pretending" (62), where the boy has to accept the reality of loss and learn to live in connectivity. This assertion posits a non-anthropocentric perspective on human-nonhuman relationships.

However, in the next two scenes, Max, who sees himself, his family, and his country as free from the threat of extinction, rubs his hands across the board angrily and smears the word *EXTINCTION*:

KHIM: Americans are very bad about understanding the concept of extinction. . . . They do not believe it will ever happen to them. . . .

MAX: You're full of shit. People don't become extinct. . . . People are smart. . . . We're not like some . . . stupid bug, who doesn't know enough to move somewhere else, and just sits there in the jungle, waiting to . . .

KHIM: Where would we go? . . .

MAX: We have the whole planet!

KHIM: Not so big now. Nineteen fifty, we have less than three billion people here. Now, more than six billion. Two thousand fifty, projected more than nine billion. You see how this goes. This is many neighbors, each using resources, each taking up space, many, many people, and more coming. You think we're going to go live on the moon, now? (45, 51)

This exchange foregrounds the necessity of thinking about the possibility of human extinction. If human and nonhuman forces are affected by each other in the era of global environmental change, humans can no longer delude themselves about being distinctive, exclusive, or masterful simply by virtue of constructs such as class, nation, gender, systems of knowledge, metaphysics, religion, or even art. The dialogue between the teacher and the student – in its scientific take on ecological relationships – offers an ecodramaturgical model. In Max's reaction to the possibility of human extinction and Khim's response, we see a dark picture of a planet that has undergone unprecedented change. As the number of people increases, humanity's ecological impact intensifies. Large-scale effects including the decline of nature's resources, species extinction, and catastrophic variation in environmental conditions may extend beyond our knowledge and bring about our own extinction. Their dialogue thus reflects Colebrook's question: "How is it that the human species, seemingly so hungry for life and dominance, has conveniently forgotten its own self-extinguishing tendencies?" (*Death of the PostHuman* 11). Considering the impacts of anthropogenic activities – our ecological footprint – on the whole planet, the play reminds us of the fact that our lives and deaths are linked to other species' extinction, and that no place or nation is exempt. By challenging the distinctiveness of the human and reconsidering the relationship between humans and nonhumans, the playwright juxtaposes different kinds of extinction and integrates Max's experience of losing his mother to cancer with nonhuman extinction and loss of community across Bolivia and Cambodia as well as with the sixth extinction more broadly. Such a view of ecological interdependency moves humans from the very centre of the world to "the place of no more pretending" (Lewis, *Song of Extinction* 62). This is the edge of "nowhere" that we currently inhabit: the all-encompassing era of anthropogenic mass extinction.

Staging a cosmopolitan performance

The concurrent unfolding of these distinct stories of death and loss resituates extinction as a site of ecological and cultural interaction, allowing an opportunity to explore the connections between the local and the global in the play. While the global conception of mass extinction is, as Khim suggests, mind-boggling and difficult to imagine because it has been abstracted and "depersonalized by data and rates of loss over time" (Arons and May, "Ecodramaturgy" 188), its juxtaposition with human mortality and the personal and regional examples of human and nonhuman extinction allows us to imagine it in a more palpable way. In their article "Everyday Ecocide, Toxic Dwelling, and the Inability to Mourn" (2020), Owain Jones, Kate Rigby, and Linda Williams argue that "despite the consistent presence of the meta-narrative of global extinction, it is in the flourishing of quite particular forms of life that their eventual demise is most keenly felt" (390). Seen in this light, the play focuses on how narratives of particular forms of extinction are "specific in their material-political-cultural-ecological unfolding and are also temporally and materially more complex and contingent than a simple sense of inevitability would suggest" (Jones et al. 390). The play's intimate exploration of this complexity and contingency contributes to our ecological awareness about the materiality of extinction in a cosmopolitan context, and to our imagination of global ecological interconnections.

The play is grounded in the materiality of micro-local scales of extinction while simultaneously engaging with the global environment. Its reshaping of the idea of extinction – from individual mourning to local and then global loss – moves the play towards a "deterritorialized" perspective. Indeed, Ellery's obsession with protecting the Bolivian forest from destruction, Lily's imaginative travel to Bolivia as she approaches her death, and Khim's acknowledgement that "I live in Cambodia my whole life when I do not see Cambodia for forty years" (Lewis, *Song of Extinction* 81) are examples of the characters' struggles to be connected with the materiality of a particular place. However, that which is intimately local about Cambodia and Bolivia is reconfigured and displaced by the shared risk of extinction, juxtaposing them with the death of a mother in yet a different part of the world and even with mass extinction on a larger environmental scale. For example, while his identity is rooted in the land of Cambodia, largely via memories, Khim's perception of the danger of an imbalanced world headed towards extinction threatens to break his bond with his homeland. On first appearance, the play seems to enact what Chaudhuri terms "geopathology," or disordered attachments to place and home (*Staging Place*). Nonetheless, the play seeks to sketch out an alternative deterritorialized framework by interrupting Khim's story with multiple narrative voices of loss. In this way, the play presents an instance of what Chaudhuri has described as

> a new kind of placement, not in any one circumscribed and clearly defined place but in the crossroads, pathways, and junctions between places. At the extreme it advocates a new kind of placelessness. . ., suggesting . . . the possibility of a *polytopianism*: placelessness not as the absence or erasure of place but as the combination of layering, one on top of another, of many different places, many distinct orders of spatiality.
>
> (Chaudhuri, *Staging Place* 138)

This condition of placelessness, achieved through an intimate interconnection with multiple spaces and alternative narratives, is evident in relation to each region. For instance, the anthropogenic destruction of biologically rich forests in Bolivia connects the region to rising global extinction rates. By reconsidering the materiality of local extinctions through a deterritorialized and placeless perspective, we might acknowledge how the play represents ecological threat on a global scale.

To get a broader perspective on the play's dramatisation of extinction in terms of its local-global relationship, I argue that what we see in the play is a disordered cosmos – rather than a unified globe-shaped image – framed by a history of extinction as exemplified in the (eco)genocide events in Bolivia and Cambodia. Understanding the play from an "eco-cosmopolitan" perspective might be problematic due to Heise's focus on the way local places "encompass the planet as a whole" (*Sense of Place* 10). Heise's concept of eco-cosmopolitanism might be understood as an aesthetic pattern based on a metaphoric vision of Google Earth's database imaginary. Here, the Earth is displayed as a whole, and various micro-locations are displayed as differing ecological and cultural contexts that can be seen when we zoom from one to the other. A broader critical perspective on the "cosmos" in eco-cosmopolitanism balances Heise's approach and helps us examine how the play assembles heterogeneous kinds of extinction among humans and nonhumans. This gives a deterritorialized view of the play, not in the sense that these inhabitants encompass the planet as a whole but in a way where, using Bruno Latour's words, their "composition" builds a "common world" that is itself "a fragile, revisable, and diverse composite material" ("Compositionist Manifesto" 474). Calling for a type of "cosmopolitanism" that is premised on the theory of "compositionism" and the idea of a "loop" instead of a "globe," Latour argues that there is no unified and all-seeing view of the globe as a sphere. In his fourth Gifford Lecture, "The Anthropocene and the Destruction of the Image of the Globe" (2015), Latour calls for a different image of the globe founded on a "complex assemblage" whose agency is distributed across a wide range of locations and human and nonhuman forces ("Facing Gaia" 83). Through the slow work of composition – a continuity that is constructed slowly and progressively to connect such apparently discontinuous examples of human and nonhuman extinction – a "common world" is built, evoking what Latour calls *kakosmos*: a dark and disordered cosmos that is not well-composed. In the play's embodiment of such a common world, "consequences overwhelm their causes" (Latour, "Compositionist Manifesto" 484) in a way that makes it impossible to flee from the horrific results of past anthropogenic activities and imagine the world as a unified whole.

In problematising the ideal idea of the globe, the play resonates with Latour's concept of the loop, which illuminates the problem of scale. According to Latour, "in spite of the illusion provided by the intoxicating manipulation of Google EarthTM, scale is the result of the number of connections between localities not the circulation through any preordained zoom from the very big to the very small" ("Facing Gaia" 93). These loops or connections come together in a vast composition process, Latour argues:

> we have to weave ourselves, to cocoon ourselves within a great many loops so that progressively, thread after thread, the knowledge of where we reside and on what we depend for our atmospheric condition can gain greater relevance and feel more urgent. This slow operation of being wrapped in successive looping strips is what it means to be 'of this Earth.'
>
> ("Facing Gaia" 95)

Observing the play's common world through this lens shows how a local extinction spins a loop, which in connection with subsequent loops and the loops of other types of anthropogenic extinction, slowly, progressively, and painfully compose earthly extinctions. Each type of human-induced extinction in Cambodia and Bolivia seems to whirl its own loop further in order to cocoon itself within the ecological cycle. For example, Bolivian deforestation spins an enormous entanglement of loops: from accelerated soil degradation and erosion, greenhouse gas emissions, and disruption of the water cycle and the nitrogen cycle to biodiversity loss and the personal trauma of Lily's death. Not only the play's theme but also its form is structured in a looping way as, through quick-fire successive or coinciding scenes, it weaves each extinction scenario thread by thread until their complex assemblages cocoon all human and nonhuman agencies – who may or may not be responsible for extinction events. Even the audience may have the feeling of being wrapped in growing numbers of crisscrossing loops of awareness that – in the Anthropocene – the consequences of what we have done are coming at us and we cannot escape. The idea that people are inevitably cocooned within the "geostory" of each loop signifies the roundedness of the Earth because constructed loops eventually come back to their point of origin, and humans come to experience the consequences of their own actions (Latour, "Facing Gaia"). In that respect, each loop in the Earth's unstable and uncertain envelope is "sensitive" to the world around it and very quick to react.[3] Living in the Anthropocene, Latour argues, calls for a sensitivity that "applies to all the agencies able to spread their loops further and to feel the consequences of what they do come back to haunt them" (96). The play asks us to be cautious, responsive, and sensitive to the entangled loops; otherwise, we are doomed. Khim tries to render Max, who insists that people will never become extinct as dinosaurs did, sensitive to the changes happening around him so that he can see where he stands within the network of multiple loops.

In a scene that vividly demonstrates the looping composition of the play's common world, Khim unexpectedly finds himself with Lily at her bedside as her condition deteriorates, helping her take an imaginary journey back to Bolivia (see fig. 4.2). In the 2009 EcoDrama Festival's workshop production held at the University of Oregon, Lily's transition between Bolivia and the hospital room was made possible by a "rolling bed/boat" (Lewis, "Pictures" n. pag.). On their way down a river in Dream Bolivia, Khim encounters radically altered landscapes that are familiar but also strange. Khim struggles between past and present, looking backward and forward, although he does not wish to distract Lily from her fantasy of natural beauty filled with showers of butterflies:

LILY: I don't know where we are. (*The two of them look around like the travellers they are.*) (LILY *paddles a little. Pushing forward.*)

KHIM: Bolivia.

LILY: Bolivia? (*The boat drifts past the broken stumps of cut-down trees. Empty, deforested land.*) (*The bank of the river becomes red. Like the ground is bleeding.*) (KHIM *glances over his shoulder. There are white bones in the red earth.*)

KHIM: Maybe Cambodia. (LILY *stops rowing.*)

LILY: You keep looking behind us. (KHIM *looks forward.*)

(Lewis, *Song of Extinction* 78–9)

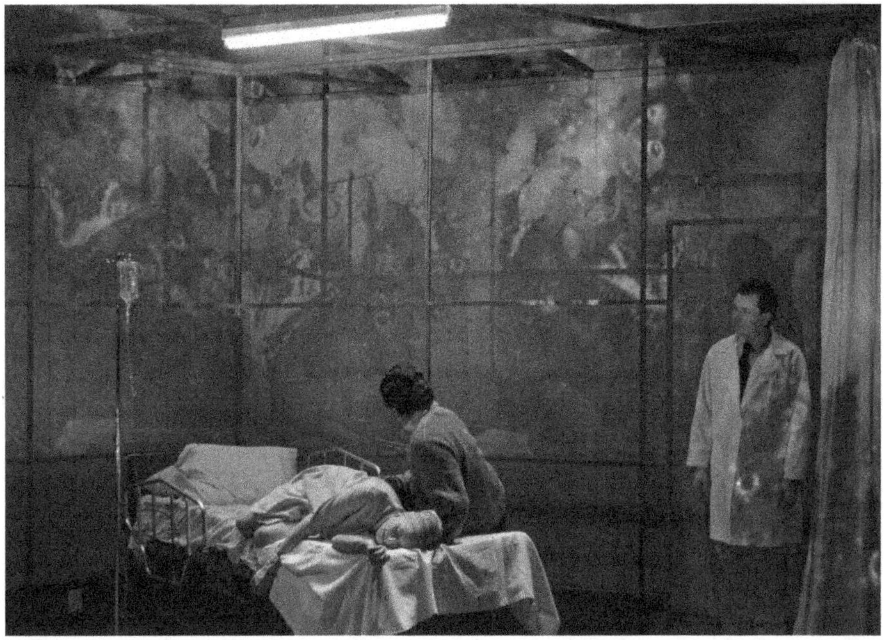

Figure 4.2 David Mura as Khim and Carla Noack as Lily. *Song of Extinction*. Theatre Latte Da, Minneapolis, 2011. Directed by Peter Rothstein. Photo credit: Michal Daniel.

Here, we witness the increased crisscrossing of diverse paths so that we may, in turn, see how they are connected. The empty and deforested land of Bolivia and the bleeding ground of Cambodia are linked to the point of conflation. In the 2009 EcoDrama production, this reciprocity was represented through the use of a red-painted floor. Extinction takes on a global meaning and loses its bond to a particular area and nation. Khim's hesitation, looking backward and forward, suggests that in order to discern what is waiting for us ahead, one needs to look back and review the past. As Latour says: "Let's stop fleeing, *break for good* with our future, turn our back, *finally*, to our past, and explore our new prospects, what lies ahead, the fate of things to come" ("Compositionist Manifesto" 486). Instead of fleeing entirely from the horror of his past and casting it as only a fragile illusion, Khim reflects on new prospects by learning from the past as an experiential foundation. The shadow of Cambodia and Bolivia – indeed, the agency of extinction – is distributed everywhere, prompting Khim to respond to Lily's question about where they are: "Place of no more pretending" (Lewis, *Song of Extinction* 79). Reflecting the pervasive complexity of the human-induced mass extinction event that the biology teacher describes to his students, the planet as a single whole is dismantled into a dark, common world composed of interconnected paths of loss (Bolivia, Cambodia, Portland), which can be intimate to or distant from our lives.

The edge of extinction and becoming

The play asks us to reconsider the interface of human and nonhuman species as well as find new ways to think about, manage, and construct knowledge about our future. Instead of taking a critical stance toward the idea of extinction or pondering it as a source of grief and mourning, the biology teacher asks his student to understand and embrace it as an inevitable part of life. According to Donna Haraway, "Mourning is about dwelling with a loss and so coming to appreciate what it means, how the world has changed, and how we must *ourselves* change and renew our relationships . . . with those countless others being driven over the edge of extinction" (*Staying with the Trouble* 38–9). In the play's final scene, the audience sees Lily's dead body in Khim's arms in the hospital room. The Bolivian jungle, to which Lily made a dreamlike journey, has receded, while the yellow butterflies that earlier fluttered around her bed have lingered. Two imaginary scenes of yellow butterflies, one in Khim's memory of his homeland and the other in Lily's dreamlike journey to Bolivia, derange any border between the two landscapes. Max and Ellery return to the hospital room and gaze at the empty bed while Khim, who has already walked over to the part of the stage where he speaks to the audience, tells us about the other side of extinction, which he thinks he should share with Max: "How do I say to him, 'Do not live in Cambodia!!'" (Lewis, *Song of Extinction* 81). This might point to a new beginning that is painfully waiting for Max's family, and for us, on the other side of distressing loss. Extinction, in its deterritorialized sense, has taught Khim dark lessons including, in Heise's words, "alienation, social uprooting, economic displacement, cultural unease, or psychological discomfort," but it may also "welcome new forms of connectivity, new choices, and a general broadening of existential horizons" (*Sense of Place* 152), which Khim intends to teach Max.

Each character experiences his or her own journey of extinction, and the playwright layers one picture of loss onto another to create a mutual entanglement of painful and dark but heartfelt and intimate stories, opening up new possibilities for thinking and forging relationships. These stories, to use Rose's expression, have the potential "to pull readers [and here audiences] into ethical proximity" ("Slowly" 9), which "takes place in a domain of entangled worlds of life and death within which we are exposed to our shared precariousness and express our responsiveness to the vulnerability and suffering of others" ("Monk Seals" 135). An example of such exploration is in the play's final moments, as the setting continuously changes back and forth. Max is in two locations at once and he moves between three timeframes: the present time; the past, when he returns from the funeral with his father, to whom he is now more conciliatory; and the future, in the classroom with Khim a month later. Max, still and disconnected, has completed his paper, which was initially supposed to be on the topic of extinction but has turned out to be about his father:

KHIM: (*reading*) His name is Ellery Forrestal. He's a biologist.
MAX: Last month, he lost everything. (MAX *glances at* ELLERY, *then back at* KHIM.)
KHIM: (*reading*) Endemism is when a species doesn't exist anywhere else. When it exists only in one place in the entire world. Last month, the subspecies of Dynastidae Strategus that my Dad found in Bolivia, that didn't exist anywhere else, and that he'd been studying for twelve years, was totally wiped out by an evil land developer. And my Mom died. (Lewis, *Song of Extinction* 82–3)

Such a horizontal view of extinction across humans and nonhumans – the death of a mother and the extinction of a beetle – changes Max's response to the connection between them. It has also altered his relationship with his father, who has lost the endangered Dynastidae Strategus and his wife, both of whom were connected to Bolivia and their past. This evolution in Max's knowledge about extinction is amplified by the dynamic power and materiality of music throughout the play (see fig. 4.3). The composer Geoffrey Pope

Figure 4.3 Dan Piering as Max. *Song of Extinction*. Theatre Latte Da, Minneapolis, 2011. Directed by Peter Rothstein. Photo credit: Michal Daniel.

provided a viola composition called "Disembarking." The viola, which is Max's only companion and is always with him, plays constantly, backed by the rising and falling sounds of a metronome. The wide range of sounds that the ticking metronome emits mirror the arrangement and derangement of notes in Max's head and everything that happens in his imagination, from his denial, confusion, distress, and grief to the evolution of his thoughts on the deeper meaning of extinction. The play's elaboration of these sounds and their dynamic agency, together with the steady beeping of a heart monitor in the hospital, indicate that change is inevitable in our life and environment, allowing us to emotionally grapple with the losses threatened by the sixth mass extinction and promote an ethical proximity with them.

Change also happens to Ellery as he acquires, through Khim, a deeper sense of what extinction means. As we watch Khim reading the paper, we can also see Ellery and Max in the car returning from Lily's funeral – Ellery has decided not to let Max go to New Jersey to live with his aunt. While Max learns to respect other species' extinction through the death of his beloved mother, Ellery realises that part of his urgent call to conserve species demands that his human relationships, namely the relationship he has with his son, evolve. Ellery comprehends that the imminent risk of extinction, not only for nonhumans but also for people, begs the question of "how we wish to live whatever time is left for the human species" (Colebrook, "Extinction" n. pag.). Ellery's new awareness of the prospect of extinction opens him up to a different sensitivity and accountability to nonhumans as well as the humans in his life.

The agency of extinction, as a global risk scenario, transforms the characters' everyday reasoning, experiential logic, and ecological thought. Max has written the rest of his paper in a foreign language – that is, the language of music. Khim describes it thus: "He calls it the 'Song of Extinction.' Maybe this is not a song about becoming extinct, though. Maybe this is a song about becoming" (Lewis, *Song of Extinction* 84). At this moment, finally, the "Song of Extinction," which Max has developed and nourished in his mind throughout the play, resonates from the stage, inspiring new ways of being sensitive to our relationship with other humans and nonhuman species in the deterritorialized and looping composition of the material world. According to Amy Nicholson's review of the play, "Lewis' smart and honest script is one small push away from collective transcendence, as we're asked to tie the threads together ourselves" in this shared environment (n. pag.). The play's narration of ecological interconnection provokes our ethical and cultural commitment, leading to awareness and action for the next generation who are unintentionally part of a global extinction event. In a broader sense, Heise argues that "[u]nderstanding the ways in which relationships to other species already form part of our self-understanding will be useful in developing the forms of multispecies justice and multispecies cosmopolitanism . . . as possible models for rethinking humans' place in what we have come to call the Anthropocene" (*Imagining Extinction* 6). This multispecies future requires us to respect, value, and personalise other species' extinction as deeply as our own, acknowledging our reciprocal involvement in the fate of the planet in the era of the Anthropocene by putting an end to our mastery.

2 Staging climate change in the deranged and intra-active space of *Sila*

In *Sila*, we witness another depiction of extinction – this time with a focus on climate change. *Sila* re-examines all the different senses of extinction considered in *Song of Extinction* in relation to climate. My discussion of climate change in the previous chapter, along with the previous analysis of Lewis's play, provides a suitable background for my ecocritical reading of *Sila*, in which two symptoms of the Anthropocene, climate change and extinction, are imbricated and affected by each other in content and form.

Sila (2015), produced by Underground Railway Theatre, premiered at Central Square Theatre in Cambridge, Massachusetts, in April 2014. The Inuktitut word "Sila," with variations such as *hila*, *hla*, *shla*, *sla*, and *tla* in local dialects, stands for the primary life force and the fundamental component of everything that exists in Inuit cosmology (Merkur 23). It can be translated in different ways, including as the universe, nature, breath, spirit, the air, the wind, the weather, common sense, intelligence, consciousness, and, as the playwright states, it can mean "[t]he thing from which everything comes and goes back" ("Perspectives on Climate Change" n. pag.). The play is set on Baffin Island in the Nunavut territory of the Canadian Arctic and features different layers of the Arctic world and climate change from the perspectives of several apparently unrelated and distinct characters whose stories converge. The characters in these rapid-fire, non-linear, interlacing storylines include Thomas and Raphaël, Canadian Coast Guard officers; Jean, a visiting climate scientist; Tulugaq, an Inuit elder; Nuliajuk, the Inuit goddess of the ocean and the underworld; Leanna, an Inuit climate change activist; her daughter, Veronica, who is a spoken word poet and teacher; Samuel, Veronica's teenage son; and a mother polar bear, who is in fact played by the same actor who plays Veronica, and her young cub. All these human and nonhuman characters are, knowingly and unknowingly, engaged in the challenging task of conserving values that are threatened by global environmental change. The play represents the ongoing instability of an entire Canadian community struggling with the complex implications of climate change, which encompass the current and future life of the local Inuit population, the polar bears, and the Arctic landscape. These themes are depicted and woven through an assemblage of several languages (including Inuktitut, Canadian French, and English), puppetry, scenery projection, and spoken word poetry as well as questions of Inuit culture and mythology, the contemporary politics of the Arctic, and the meaning of Sila.

The play is deeply grounded in an ecological context and grows out of Bilodeau's environmental activism. *Sila* is the first play in her Arctic Cycle project and is followed by *Forward* (2017). Bilodeau, originally from Quebec but now based in the New York city, is the artistic director of this project, which involves storytelling and live performance of a series of eight plays to dramatise the impact of climate change in the eight countries of the Arctic: the United States, Canada, Greenland, Iceland, Sweden, Norway, Finland, and Russia. The project is built on the collaborative efforts of artists across cultural, geographic, and disciplinary divides, including earth scientists, social scientists, community members, and educators. Bilodeau was inspired to start the project after her first trip to the Arctic in 2007, where she saw Alaska's melting glaciers and the impacts of a warming climate. The research about the environmental transformations taking place in the high north started with *Sila*. The process of writing the play included the playwright's extensive travel in the Canadian Arctic – Baffin Island in the territory of Nunavut – and her interviews and interactions with local people, community leaders, activists, politicians, coastguards, and climate scientists – including NASA's Gavin Schmidt – who work and live in the region (Yeo n. pag.). The role of Indigenous people has especially been important to Bilodeau's play because representing their cultural practices, knowledge, and experience – both traditional and contemporary – contributes to her critique of Eurocentric conventions and colonial hierarchies in the past and the present. Further, the playwright's focus on the interaction between Inuit culture, the Arctic environment, and white settler colonial culture in the context of planetary changes contributes to dramatising the challenges of climate change across personal, local, and global scales. It is also worth noting that the Arctic Cycle has also expanded beyond Bilodeau's plays, encouraging other initiatives in response to the challenges of climate change through the medium of art. Bilodeau is also the co-founder of the blog and international network "Artists and Climate Change" and a co-organiser of Climate Change Theatre Action (CCTA). Climate Change Theatre Action is a worldwide series of readings and performances of short climate change plays presented biennially during the United Nations' Conference of the Parties (COP).

The play is centrally concerned with how the Arctic landscape, Inuit community, and contemporary Arctic policy respond to climate change and how these responses contradict and overlap with one another. The Nunavut territory has been at the forefront of national and international politics as a particularly vulnerable site of climate change. Inuit communities have inhabited the Canadian Arctic for at least four thousand years. This very long history has anchored their culture deeply in the climatic and environmental conditions of the Arctic. The rapid rate of change in this environment, partly due to Canada's varying political responses to climate change from the 1990s to the present, has profoundly affected the region and its human and nonhuman inhabitants. According to Loren R. Cass, while Canada – despite the objections of the US – was "at the forefront of efforts to address climate change and was widely viewed at home and abroad as a leading state in the diplomatic response," its support for the 1990–1992 United Nations Framework Convention on Climate Change and the 1995–1997 Kyoto Protocol was largely "symbolic" (20–1). Following the ratification of the Kyoto Protocol, Canada made a commitment that by 2012 it would meet carbon emission targets set at

6 per cent below 1990 levels. However, Canada fell short of this commitment and formally withdrew from the Kyoto Protocol in 2012. Though Canada has committed to reducing its greenhouse gas emissions to 30 per cent below 2005 levels by 2030 under the 2015 Paris agreement, Cass's argument that the country has affirmed "emergent international norms without being immediately forced to change domestic practices" still stands (26). Canada's ineffectual climate policy has disproportionately threatened the people, nonhuman species, and environment of Nunavut, which is among the places most vulnerable to the changing climate.

According to *The Arctic Climate Impact Assessment* (ACIA), four reasons compel scientists to examine the impacts of changes in climate across the Arctic region (Huntington et al. 3). First, the Arctic has experienced the effects of global warming more intensely than other regions in recent decades (3). Second, the warming trend of recent decades is predicted to cause ongoing changes in Arctic seasonal patterns and their consequences, including accelerated melting of glaciers, extensive retreating and thinning of sea ice, and thawing of permafrost. Each trend is accompanied by impacts on biodiversity and rates of extinction of nonhuman species, such as ringed seals, caribou, various species of whales, and polar bears (3). For example, the dramatic shrinking of the sea-ice habitat has posed a serious threat to polar bears, which are entirely dependent on the frozen ocean for hunting (mainly ringed seals). Third, global warming continues to escalate, affecting the economic activities, subsistence livelihoods, culture, lifestyle, safety, and health of the Indigenous communities – who, in fact, contribute least to climate change. The Inuit people's land and culture are adversely affected by climate change caused by activities outside the region, including southern infrastructural developments and mining explorations. Fourth, it is difficult to draw restrictive boundaries for the Arctic because it is "an important part of the global climate system" that "both affects and is affected by global climate change" (4). Variations in Arctic ecosystems may have enormous impacts on local and global climates.

A shortage of adaptive capacity to accelerating environmental change not only affects Arctic residents' subsistence activities but also has implications for their identity, traditional ecological knowledge, customs, values, and experiences across current and future generations. The intersection of culture and climate is evident as the foundation of the ancient culture, which has been deeply rooted in the landscape, now trembles in the rapidly melting sea ice. Cultural values and narratives shape our understanding of extinction and climate change as vivid signals of global ecological crisis and scientific and political challenges. In the context of increasing concern about global environmental crisis and the Anthropocene, Heise argues that issues of biodiversity loss and species endangerment are closely related to cultural "anxieties over the consequences of modernization and colonization" (*Imagining Extinction* 19). For Bilodeau, theater offers an opportunity to dramatise and shape the stories we tell about human relation to environmental transformations. For example, in "A Playwright's Journey to the Canadian Arctic" (2016), Bilodeau describes her research trip to Baffin Island where she discovered "a vibrant culture struggling to build an economy and catch up with a rapidly changing world that doesn't have much patience for those who lag behind" (n. pag.). Many of the scientific, economic, environmental, and cultural challenges faced by the Inuit people as well as nonhuman Arctic species find their way into her play *Sila* (Bilodeau, "A Playwright's Journey"). While the play is fact-based, it creates a fictional, dramatic, and material representation of the stories of human and nonhuman species affected by climate change.

A stage of derangements of scale

The play begins with two contradictory epigraphs describing the Arctic. The first is from the book *Arctic Dreams* (1986) by nature writer Barry Lopez, who provides a coherent and transcendent depiction of the spirit of the landscape that is, in fact, the landscape of the past. The second is by Sheila Watt-Cloutier, an Inuit climate change activist, who announces an imbalance in the landscape of the present. These two pictures align with two distinct representations of the Arctic that the playwright intends to dramatise: the Arctic of the Inuit, which is "warm, raw, and fiercely alive," inhabited by spirits, gods, immortals, dance, and music, and the Arctic of the Southerners or "the new North," which is "cold, mystical in its foreignness, and rarefied" (Bilodeau, *Sila* 5). With melting glaciers and endangered polar bears, the latter landscape embodies "the excesses of modern life" and anthropogenic disturbance (Bilodeau, "Artistic Contribution" n. pag.). To explore these disparate images of the Arctic region, I turn to the first scene, in which Leanna, an Inuit climate change activist, describes the landscape from which she comes:

> I come from a place of barren landscapes and infinite skies. I come from a place of rugged mountains, imperial glaciers and tundra-covered permafrost. . . . I come from a place where the only crowds are air, sea and land creatures that gather each year by the thousands. . . . I come from a people who have kept accounts of the early days when the world was rich and urgent . . . When spirits roamed the land like polar bears and muskoxen and caribou. . . . I come from a land whose wisdom reminds us of our humanity. This place I come from we call Nunavut. It means 'Our Land' . . . where we, Inuit, have thrived for more than four thousand years. . . . It's where we nurture our knowledge of who we are. But Nunavut, our land, is only as rich as it is cold. And today, most of it is melting.
> (*Sila* 12)

In Leanna's description of the magnificence of the deep white of the Arctic landscape, we can see the depth of connection between the land and its human and nonhuman inhabitants. Such practical intimacy linking people, mountains, air, sky, sea, glaciers, permafrost, and animals – each with its own material story – is depicted not only geographically but also socially and culturally. The land's intelligence is anchored in the social and cultural life of its inhabitants in their consciousness, imagination, emotion, and shared humanity. However, this richness and intimacy grow strange as the glaciers melt, possibly causing the extinction of nonhuman species.

The play's 2014 premiere at the Underground Railway Theatre in Massachusetts, with Megan Sandberg-Zakian directing, materialised the vastness and whiteness of the Arctic tundra in a simple and pristine way by deploying layers of white curtains stretching from ceiling to floor (see fig. 4.4). According to Sandberg-Zakian, such a beautiful poetic space was needed for "the impact of loss to be deeply felt on a large scale" (i). The fluid curtains of sheer fabric made a large screen for the projection of underwater and slow-moving masses of ice as well as celestial scenes, the Northern lights, mythological images, and Inuit stories recounted via the stars of the constellation Orion and shadow puppets. For the director, the play's themes of interconnectedness and isolation were also explored through this use of fluid white space that "could merge or consume our locations of tundra, living room, Coast Guard office, and ocean" (i). These different locations were brought together and continually reshaped by the impact of climate change, which is common to all yet experienced differently by each. In another production of the play by the University of New Hampshire and its Sustainability Institute in 2014, set designer Szu-Feng Chen – who is also

Figure 4.4 Layers of white curtains stretching from ceiling to floor. *Sila*. Underground Railway Theatre Company, Cambridge, MA, 2014. Directed by Megan Sandberg-Zakian. Set design by Szu-Feng Chen. Photo by Szu-Feng Chen.

the designer of the Underground Railway Theatre's production – used almost two thousand recycled plastic bottles to separate the human interior space from the soft fabric covering the Arctic landscape. According to Chen, although the blue colour of the wall reminds us of the frozen ice and clean home of the polar bears, it represents the human world because "the man-made, artificial quality in the icy plastic [is] ironic to the fact of disappearing ice land" (n. pag.). While the wall projected the theme of sustainability in theatrical design, it served as a background for displaying varying stories of the Arctic world.

Leanna's introduction of a poetic and sublime Arctic is shattered by the scene in which Thomas, a Canadian Coast Guard officer, attempts to convince Jean, a French-Canadian scientist and climate researcher, to conduct an environmental assessment for an industrial plan to set up a deep-water port for oil exploration. Jean, however, does not accept the proposal because he believes it intrudes on Inuit traditional hunting and fishing rights and disturbs marine areas and species. He responds to Thomas' suggestion thus: "the environmental assessment is just a marketing ploy. You know that. They'll look at it and drill anyway. . . . I learned my lesson: science and politics don't mix" (Bilodeau, *Sila* 14). Thomas, however, is determined to assert control of the region and its natural resources by building infrastructure and a seaport, even at the expense of land degradation: "There's too much money at stake. If we wanna maintain sovereignty over our Arctic territory, we need to establish a strong presence" (16). For him, "[t]aking the lead in exploiting . . . resources is one way to assert sovereignty" before other industrialised countries such as Germany, Russia, and the US drill and conquer the land (16). In order to grant the project validity, he asks for Jean's support, forcing him to extend his contract with the oil company: "Having you, a CANADIAN and one of our most prominent scientists, doing research is another. It shows that we're interested. It shows that we care" (16). Science is, for Thomas, a way to justify drilling for oil and veil the political and economic imperatives behind it and other ways of exploiting the Arctic environment. Nevertheless, this environmental assessment is, in reality, a threat to local groups and the global environment. Thomas also asks Leanna to publicly support the project, hence giving it credibility and persuading the public of its necessity. Leanna argues: "I can't in one breath talk about protecting the Arctic and, in the other, lay out a plan for dredging a channel right in the middle of it!" (20). Her environmental activism, concerned as it is with protecting the land from the risk of oil spills and carbon emissions, holds Leanna back from involvement in Thomas's project. The changes noted in her description of Nunavut – from cultural, national, and ecological rootedness in a balanced landscape of ice to a warmer and less lively place, where ice glaciers are melting away – position the local environment in relation to large-scale ecological threats caused by anthropogenic activities.

The play deals with the ways in which large-scale climate change disrupts basic premises about knowledge and distinct patterns of action by confusing the boundaries between science and politics and simultaneously engaging different scales of comprehension. In particular, actions that have been traditionally considered progressive in terms of economic growth, national security, and environmental protection are problematised. The following exchange between Thomas and Raphaël, another Canadian Coast Guard officer, demonstrates some aspects of this issue:

THOMAS: The effects of seismic surveys on marine life have been studied for decades. There's no proof that they do any harm.
RAPHAËL: Sure. Firing air guns into the water every ten to fifteen seconds is a totally benign activity. And getting money from Natural Resources Canada doesn't at all mean that this vessel is looking for oil.
THOMAS: Their mandate is to map the sea floor. They're not responsible for what people do with that information. . . . The data the *Polaria* will provide can help us get more resources.
RAPHAËL: New icebreakers?
THOMAS: Nevermind icebreakers. What I'm going after is a port. . . . A state-of-the-art, environmentally responsible, military and civilian loading and refueling station.
RAPHAËL: And where are you gonna put that port? In Hyperborea? (47–8)

For Thomas, the data gathered by the *Polaria*, a German scientific ship, through seismic testing in Baffin Bay can help to exploit more resources to create industrial centres, seaports, and infrastructure. Thomas tries to convince the Canadian government and Inuit self-governing bodies to build a seaport that he promises can be "environmentally responsible" (48). However, for Raphaël, Thomas's illusory seaport, which supposedly holds the key to Canada's political and economic supremacy without harming the environment and marine life, only exists in Hyperborea – an imaginary place in Greek mythology that is warm and sunny as it lies beyond the cold breath of Boreas, the north wind. What seem to be rational actions for modern Western societies – namely economic policies that elevate prosperity – need to be reassessed in terms of their destructive ecological impact on the individual, local, and global scales. Challenging the famous slogan, "think globally, act locally," Timothy Clark argues that eco-global issues such as climate change subvert the possibility of acting only locally, so that, paradoxically, "one cannot *only* act locally" because "any action affects the whole world, however minutely" (*Value of Ecocriticism* 39). Likewise, the play shows that it is difficult to perceive the impact of global environmental changes or the possible consequences of our anthropogenic activities from a personal perspective, as they are extended across time and space, embracing both humans and nonhumans. Clark also argues that the Anthropocene and its symptoms, such as climate change and extinction, are themselves emergent "scale effects." These scale effects are so complicated and deranged that it is difficult to comprehend the magnitude of impact:

> Scale effects in relation to climate change are confusing because they take the easy, daily equations of moral and political accounting and drop into them both a zero and an infinity: the greater the number of people engaged

> in modern forms of consumption then the less the relative influence or responsibility of each but the worse the cumulative impact of their insignificance.
>
> ("Scale" 150)

In such contingent relations, "what is self-evident or rational at one scale may well be destructive or unjust at another" (Clark, "Scale" 151). The play involves such "derangements of scale," in which anthropogenic activities – including transformations of the landscape and resource extraction, which unavoidably accompany increasing human consumption of fossil fuels – are incommensurable with the global effect or actual scale of cumulative environmental degradations such as climate change.

Against these scenes of political and scientific debate about whether the Arctic's resources should be subject to exploitation as the sea ice melts, a different imaginary, that of Inuit ecology and culture, is enacted by Leanna's daughter and the polar bears. In one scene, we see spoken word poet Veronica performing a poem about Inuit hopes and dreams of the land. Through the deeply entrenched culture of oral storytelling, she depicts a material and intimate interchange between the living land and its inhabitants. The play is also narrated from the perspective of Mama Bear and her young cub, whose Inuktitut names are Anaanaa and Paniapik respectively. They appear on stage in the form of life-size material puppets or are represented, in some scenes, as shadow puppets (as in the 2014 Underground Railway Theatre and University of New Hampshire productions which use the same puppets) (see fig. 4.5). The polar bears wait by a seal breathing hole for their meal and wander over the floating ice sheets, which are progressively disintegrating. In the meantime, they track down stars of the constellation Orion and recount mythological stories of a time when "strange powers were shared among animals, people and the land, when all creatures spoke the same tongue" and when *Nanurjuk*, a star in the constellation Orion that has the spirit of a polar bear, roamed the earth (Bilodeau, *Sila* 19). However, this ancient version of the North, a land of eternal harmony – which the play's Inuit communities and polar bears imagine, in the manner of sixteenth-century mapmakers, as "a circular continent surrounding a polar sea with a single high mountain in the center" (Bilodeau, "Artistic Contribution" n. pag.) – no longer exists due to the impact of climate change.

Although the play takes place in the territory of Nunavut, the action is built upon a landscape that is shifting due to climate change. The Baffin Island region in the play, as Arons and May describe it, acts as a kind of *frontera* that is fluid and "permeable to Inuit, Western (Canadian-American-German), scientific, governmental, animal, and mythic/spiritual realms" ("Ecodramaturgy" 191). Its permeability disrupts any distinction between local and global geographies made by ethnic, civic, and national identities (191). In this way, according to Heise, the cultural meaning of the polar bear is "not keyed to a particular community so much as to the planet as a whole" (*Imagining Extinction* 241) because global warming is transforming the Arctic environment so that polar bears drift

Figure 4.5 *Sila*. Underground Railway Theatre Company, Cambridge, MA, 2014. Directed by Megan Sandberg-Zakian. Puppet design by David Fichter. Photo by David Fichter.

on shrinking ice floes. Hence, I argue that the play contributes to this global way of imagining polar bears as an endangered species that embodies the threat of climate change. To frame the play within a bioregional perspective or a particular historical period would put it on the wrong scale. The reality of climate change makes us rethink the landscape of the Arctic as a radically altered and displaced area embedded materially, ecologically, and culturally in the current degraded state of the planet – and being made worse by international exploitation. Returning to Leanna's first description of the Arctic, the landscape is depicted as a place of "rugged mountains, imperial glaciers and tundra-covered permafrost" where the seemingly eternal icy skyscrapers stand proudly as immoveable obstacles that shape national barriers and project spiritual sublimity and cultural wisdom (Bilodeau, *Sila* 12). However, the melting of glaciers and the reduction in the ice's thickness are a few of many phenomena that subject the landscape to continuous change, both literally and figuratively. This change is powerfully represented on stage through the young cub's dramatic fall through the thinning polar ice.

In such a displaced environment, "land" is overlaid by "climate,"[4] producing a new consciousness built on a decentralised narrative voice and a redefinition of human spatiotemporal scales. After spending three weeks in the Baffin Island region exploring its human and nonhuman inhabitants and talking to scientists,

Inuit activists, and elders, the playwright changed her initial plan to create a traditional storyline. For Bilodeau, a narrative structure driven by the perspective of a singular character on a particular subject felt too simplistic and inappropriate in the Anthropocene epoch ("New Aesthetic" n. pag.). She writes:

> And since greenhouse gases permeate the global atmosphere, there is no way to address the problem locally, without taking the entire system into account. This is particularly true in the Arctic where the Inuit population is being disproportionately affected by warming temperatures, yet produces very little carbon emissions. The Inuit cannot, on their own, protect their home because the forces threatening it are coming from places that are not under their jurisdiction.
>
> (n. pag.)

The playwright's description of the global scale of climate change highlights a huge shift in the human perception of time and space. Interference with natural processes, infrastructural developments, and, above all, greenhouse gas emissions are not palpable as singular phenomena. Rather, these anthropogenic activities have effects through their relationship with manifold kinds of environmental forces, which exert their transformative power over a vast scale of time and space. Considering the global scale of human carbon footprints, Clark asserts:

> If it were just a matter of my own emissions there would be no controversy. The size of my carbon footprint is of no interest or significance in itself except in relation to the incalculable effect of there being so many millions of other footprints over an uncertain timescale, which is something beyond my individual phenomenological horizon entirely.
>
> ("Phenomenology" 286)

Clark's concept of derangements of scale acknowledges the way our anthropogenic actions have themselves been decentred by our inability to identify our responsibility for their implications over the course of recorded history or in any single instance. The play shows how the relationship between action and consequence becomes disoriented and how anthropogenic actions are dislocated in that they produce outcomes in unprecedented, delayed, and disparately and unequally disseminated ways.

The anthropogenic manipulation of the carbon cycle alters familiar modes of conceptualisation, assigning strange meanings to natural history as well as to human history and culture. The Arctic as described by Leanna, with its only "crowds" those of creatures of the air, sea, and land, has undergone a radical mutation due to geological change induced by material infrastructure and human activities. In this way, the Arctic landscape exemplifies a "storied matter" (Iovino and Oppermann, 1) narrating its own agency by affecting the global scale, not in a normal and harmless way, but in a destructive one. The narrative potentiality of the Arctic is composed of the incongruent but intertwined human and nonhuman perspectives and voices as each plays out its own story. Of course, mapping the scale of such narrativity runs into difficulties of its own, as Clark explains: "the difficulty of conceptualizing a politics of climate change may be precisely that of having to think 'everything at once'. The overall force is of an implosion of scales, implicating seemingly trivial or small actions with enormous stakes while intellectual boundaries and lines of demarcation fold in upon each other" ("Scale" 152). Such a process of thinking "everything at once" in this new experience of global change illustrates an illusory hierarchy of scale, and, in fact, a definitive feature of ecocritical reading and environmental thinking in the Anthropocene epoch. As such, the play thoroughly dislocates and redefines familiar concepts of rationality, responsibility, action, and equity. This exploration calls for various perspectives, not a human-centred one, to measure the scope and complexity of the cumulative impact of climate change. I now consider how, through a horizontal frame of relationships, the playwright integrates different narratives so that both humans and nonhumans convey their distinct agencies, stories, and tragic losses.

A stage of intra-active performativity

It is hard to know the scale – whether that of humans, polar bears, the Arctic, fossil fuels, or the Anthropocene – at which we see the play's representation of climate change, because climate change involves manifold stories and leads to derangements of scale. The play's juxtaposition of different tragic worlds of climate disruption is narrated through the predicaments of various participants in a community: Samuel, Veronica's teenage son; a young polar bear cub; and Jean, who falls into the sea, where he encounters Nuliajuk, Inuit goddess of the ocean and the underworld. First, I focus on two distinct stories of devastating loss – for Leanna's family and the polar bear – that are presented sequentially. Leanna warns the government and its bureaucracy of the consequences of oil drilling, petitioning them about the necessity of protecting Inuit culture and Inuit rights on Baffin Island. In return, she receives a letter from the Commission denying her request despite the scientific research she has painstakingly compiled. Leanna is concerned with preventing roads and houses from sinking and teaching Inuit children the character-building skills their ancestors traditionally learned. Nonetheless, she does not consider that climate change could have an even more immediate impact – one soon embodied in the death of her grandson, Samuel. Veronica, meanwhile, is struggling with her troubled teenage son who suffers from depression principally caused by the loss of land and culture. Samuel goes on to take his own life. Upon hearing of her son's death, Veronica remains glued to the spot, speechless.

This scene is followed by the death of a polar bear. By convention, only humans can be tragic subjects, but this play's ecodramaturgical shift shows that animal loss can also be the emotional focus. The polar bear Mama and her cub are stranded away from feeding sources on floating Arctic ice. As they become weak with hunger, they hear "[e]erie sounds of cracking sea ice followed by the howling of a rising wind" (Bilodeau, *Sila* 53). The cub, Paniapik, is wondering if humans have broken the ice and are trying to kill them, but her mother, Anaanaa, explains that "[t]he ice broke because Nuliajuk is angry. And Nuliajuk is angry because humans have angered her" (55). Human neglect and abuse of the natural environment make Nuliajuk angry and vengeful. In the next scene, the cub perishes when the melting ice floes crack and turn over, throwing her into the water. She drowns despite her mother's desperate struggle to stop her daughter's body from being dragged beneath the sea:

> *The DAUGHTER's panting gradually slows down as the Earth's breathing rises. MAMA swims. Soon, the DAUGHTER and the Earth breathe in unison, their breathing rising up and down with the swell of the ocean. . . . A moment of perfect harmony. Then as the DAUGHTER's breath starts to fade, long strands of something – plants perhaps – emerge from the water. . . . The strands wrap themselves around the DAUGHTER. . . . She sinks into the water as the sea swallows her breath in one swooping motion.*
>
> (60–1)

This scene of tragic loss begins with a moment of complete harmony between Paniapik's breath and that of the Earth. For the Inuit, one of the traditional meanings of Sila is "breath-soul" or *inua*, which refers to an intimate material power dwelling within each living organism, giving it life and awareness (Merkur). The animal's breath-soul is, however, part of the broader breath-soul of the earth, which is composed of interconnected individual *inua* (Merkur). The Earth's climate, this cosmic breath, is, of course, disrupted by human-induced ecological changes, making the sea, represented by the mythical figure of Nuliajuk, angry and vengeful and consequently affecting all the other individual breath-souls. After a desperate tug of war between Anaanaa and the deadly sea, the cub's body is enveloped by the waves and sinks to the bottom. Sensing that she will not see her cub again, Anaanaa shouts aloud her name, lets out long desperate wails, and then cries out softly in quiet despair: "*Paniapik*" (Bilodeau, *Sila* 58).

In the age of climate change, the fate of the polar bear family echoes that of the human family. The picture of common breath described above brings the traditional knowledge of Sila into the shared predicament of climate change as a communal challenge for all species. In this context, the loss of a nonhuman being is treated as gravely as the loss of a human loved one, prompting us to reflect on Colebrook's question: "How might we imagine a world without . . . the centered points of view of sensing and world-oriented beings?" (*Death of the PostHuman* 23). This question puts an end to human exceptionalism and calls for decentred relationships within the world (20). In giving both the polar bear mother and Veronica the same voice, and in representing

Veronica expressing her silent grief through the wails of the polar bear, the play enacts an exchange of the breath that is common to humans and nonhumans, though each participates in this relationship with different kinds of agency and particular capacities, senses, and experiences. For Yupik Native law student Meghan Siġvanna Topkok, who played the Daughter bear in the play's production at the University of Oregon in 2015, "kinship with animals is central to indigenous Arctic identities. . . . 'It's not a metaphor'" (May, "Radical Empathy" n. pag.). May, who directed this production, described the polar bears as "rehearsed not as objects (puppets), nor even as individual characters, but as possibilities of becoming. 'Becoming polar bear' but never 'acting' bear, sharing breath and continuous movement, an ensemble shaped and re-shaped, feeling into the question of kinship" (n. pag.). In sync with this material approach to the representation of nonhumans and the performative agency of the Arctic environment, I argue that this theatrical performance plays a pivotal role in communicating an ecological and material multispecies worlding, in which humans, nonhumans, and environmental forces are entangled with and affected by each other, all in the process of becoming in a rapidly changing world. This stage of shared fate opens up a space of performance that Chaudhuri calls "the theatre of species." For Chaudhuri, the theatre of species, as "an emergent performance practice of our times," "brings the resources of performance to bear on what is arguably the most urgent task facing our species: to understand, so as to transform, our modes of habitation in a world we share intimately with millions of other species" ("Silence of the Polar Bears" 50). Given this intimately multispecies theatrical space, I add that the pairing of the polar bear mother and her cub with Veronica and her son introduces a new approach in the theatre of the Anthropocene, disturbing the familiar idea of subjectivity and the anthropocentric form of performance.

The conjuncture of these two losses implies that not only the polar bears – a species facing a high risk of extinction in the wild – but also the vulnerable members of our own species and, in particular, younger generations, are victimised by climate change, which disrupts Inuits' sense of place and four-thousand-year-old culture and lifestyle. In a way that parallels *Song of Extinction*, this tragedy causes a shift in Leanna's approach to climate change, which for her changes from an abstract concept in a research project to an intensely personal, intimate, and material issue. In her second monologue, and after losing her grandson to suicide, Leanna expresses a transformed awareness of the Arctic landscape threatened by climate change. In her speech before the Inter-American Commission on Human Rights, she reproaches industrialised countries for not taking action to reduce their emissions and for disregarding basic Inuit rights to life, health, and culture. She argues thus:

> So you may tell me that the world's economic survival is more important than the well-being of a small Arctic nation. You may tell me that anxiety and fear and depression are a matter of personal choice, not of environmental stewardship. You may tell me that drug abuse and . . . teenage

> suicide . . . are by no means a sign of degradation of the Arctic. But I am here to tell you otherwise. . . . The real issue is not . . . climate change. The real issue is that we have lost part of our humanity. We have lost our capacity to care . . . But unless we open our hearts and embrace not just people we love, but people who are not yet even born, we will never value our species enough to make sure it survives.
>
> (Bilodeau, *Sila* 63–4)

Leanna's speech cogently embodies the intensification of human-driven alteration of the physical world, which has radically changed our mode of thinking about the future, reconceptualised the notion of humanity beyond its own borders, and, ironically, facilitated our extinction. To this end, any attempt to externalise climate ignores the fact that human life is already enmeshed in its materiality and performativity. By revealing the performative capacity of climate change and moving it from the periphery to a central position where it is embedded in the present and future life of humans and nonhumans, Leanna puts into question the mastery of the human species and calls the audience to care about the environment as well as future generations.

The play also contains a third plotline, one that involves Jean. His story provides a complement to Leanna's monologue about the need to take responsibility for and respond to the changed environment in a visceral way. Jean, who refuses Thomas's proposal, is stranded on Baffin Island without receiving any of the assistance essential to his geological modelling and to collecting data on the multi-year ice collapse, such as a helicopter. While Jean waits, he hires Tulugaq, an Inuit elder, as his guide. Tulugaq provides him with a skidoo, a drill, and other instruments for preparing a model to justify the funds he was granted for his trip; he also shows Jean the safest route and trains him to watch for polar bears that may appear on the ice. The Inuit elder takes traditional knowledge to heart by way of a deep attunement with the natural environment. Conversely, Jean is concerned only about geometric abstractions, numbers, and models. For Tulugaq, the Arctic cannot be understood simply with numbers and scientific measurements; rather, it is historically, culturally, artistically, and materially composed of stories such as those of the *aqsarniit* (Northern lights or Aurora Borealis), the weather, polar bears, and humans, among many others. While Jean focuses on his scientific analysis of Arctic sea ice conditions, the stories are alive and dynamic for Tulugaq, and they need to be observed, felt, and experienced with head, heart, hands, and feet: "Numbers are not enough. We need stories" (52). Based on this Inuit philosophy and the Indigenous traditions of performing stories about the landscape, the agency of climate change needs to be conveyed not only through the cold logic of numbers and metrics but also through emotional, social, cultural, and ecological responses. According to Rob Nixon, "to measure and project the human fingerprint in the fossil record," we need stories to make "visceral those tumultuous geologic processes that now happen on human time scales" ("Anthropocene" n. pag.). Heise also argues that the global ecological disasters of biodiversity loss and extinction are not, "in the end, just a matter of science, but also and mainly one of histories, cultures, and values" (*Imagining Extinction* 237). In the play, although Tulugaq's emphasis on Arctic stories is underpinned by Inuit culture, traditional knowledge, and harmony with nature, retelling them in connection with climate change also requires a material rather than a theoretical perspective: the changed environment is composed of the ever-forming narratives of "storied matter" while their entanglements shape a multispecies planetary ethics.

The play's staging of an agential landscape that offers a broader awareness of ethical and practical care for the environment is exemplified in Jean's fall through the ice and the subsequent danger of Nuliajuk. As soon as Jean prepares to drill holes and deploy moorings, he encounters the unconscious, starving Mama Bear underneath the ice. All of a sudden, Mama opens her eyes, props herself up, and growls, while the ice cracks. There is an audible splash and Jean falls into the Arctic Ocean. Jean's story contributes to a reconception of the human characters' connection to nonhuman animals via their shared experience of extinction. This connection is, more broadly, depicted in relation to other nonhuman forces such as the sea and climate. The play thus expands Chaudhuri's theatre of species because its diffusion of agency does not stop at animals. The mythical dimension of Inuit life is involved when, in the underwater world, Jean encounters a larger-than-life Nuliajuk with a wild mane of plant-like strands, in which various sea creatures, including the cub, are trapped. As Jean gathers his strength and tries to push himself up to the surface, the sea animals grab him while Nuliajuk wraps strands of her dark hair around him, cocooning him:

JEAN: No. Let me go. LET ME GO! . . .

NULIAJUK: To fight. Bird-spirit. Spirit shaken. Briefly. Then spirit's power grew. Dark. Heavy. Overwhelming. My father. Threw me. Overboard. Threw me. Into the icy water. I gripped. The kayak. Cut off. My fingers. . . . Here. At the bottom. Of the sea. I was reborn. Now the ocean. Is my dominion. Here. I can sense. The weakness. Of all. Humanity.

JEAN: But what does any of this have to do with –

NULIAJUK: Weak! *Her anger triggers a cacophony of frightened SEA ANIMAL responses*. . . . Like my father. Don't protect me. At a time. When. I need it most.

JEAN: Then I will I'll protect you I promise. . . .

NULIAJUK: Prove.

JEAN: Prove? Prove how? I mean I – I –

NULIAJUK: AAAAAAAAAAARGH!!! . . . Weak.

JEAN: No please . . . Noooooooooo!

DAUGHTER & SEA ANIMALS: Help me . . . Help me . . . *The phrase echoes endlessly while JEAN is swallowed into the cocoon and everything recedes into a dark blue mystery*. (Bilodeau, *Sila* 84–7)

While Nuliajuk ensnares Jean in her plant-like strands, she narrates in broken sentences the story of her past life. In a version of this Inuit myth, Nuliajuk (Sedna) was tricked into marriage with a fulmar, a kind of seabird, disguised as a man. According to Pamela R. Stern, Nuliajuk's father tried to bring his daughter back home in his kayak, but the fulmar "beat his wings so furiously that he raised a storm on the ocean that threatened to swamp the tiny boat" (2). Finally, the frightened father threw his daughter into the sea to save himself. However, Nuliajuk "grabbed onto the side of the umiak and tried to pull herself back into the boat" (Stern 2). To stop her, the father cut off her fingers, which fell into the sea and transformed into sea creatures including seals, walruses, and whales. Nuliajuk slipped under the water where she became the goddess of the sea and a host of marine animals. However, she was quite miserable because, without fingers, she could not comb her hair, and sea creatures became tangled in it (Stern 3). And now Jean is among the creatures trapped in her hair (see fig. 4.6 & 4.7). Nuliajuk's anger also manifests itself through the loss of other human life. The long strands of her hair cause storms and capture the expedition boat of a group of German scientists, bringing them to their deaths. Redolent of Caryl Churchill's *The Skriker*, Jean's journey to the underworld depicts the material environment in the action of a performance where agency is not solely a human feature but distributed across a field of human and nonhuman bodies.

Figure 4.6 *Sila*. Underground Railway Theatre Company, Cambridge, MA, 2014. Directed by Megan Sandberg-Zakian. Puppet design by David Fichter. Photo by David Fichter.

Figure 4.7 *Sila*. Underground Railway Theatre Company, Cambridge, MA, 2014. Directed by Megan Sandberg-Zakian. Puppet design by David Fichter. Photo by David Fichter.

The play's representation of the underwater world brings the materiality of the nonhuman world and unseen forces into the forefront of analysis, as they are destructively transformed by human manipulation of the earth's ecosystems. Nuliajuk, with her dark plant-like strands of hair, embodies the materiality and performativity of the Arctic Ocean as she is reconfigured into a threatening hybrid agent in which seals, various sea creatures, the cub, and Jean are tangled. This figure acts in a "tentacular" way and can be understood as an instance of Haraway's "tentacular thinking" (*Staying with the Trouble* 32). Contrary to anthropocentric modes of thought, tentacularity emphasises multiple, criss-crossing networks of lines, in harmony with Deleuze and Guattari's concept of the rhizomatic. According to Haraway, "Tentacularity is about life lived along lines – and such a wealth of lines – not at points, not in spheres. 'The inhabitants of the world, creatures of all kinds, human and non-human, . . .' are like 'a series of interlaced trails'" (32). As an ecodramaturgical feature, the composite figure of Nuliajuk and her material tentacular becomings and intrinsic performativity remind us of the fact that the universe of materiality is a complex network in which human and nonhuman forces are inextricably entangled.

The idea of tentacular thinking contributes to identifying this hybrid entity, Nuliajuk, as an embodiment of Karen Barad's call for a new materialist reconceptualisation of performativity. The performativity that Barad proposes is, in fact, an elaboration of "materialist and posthumanist reworkings of the notion of performativity" ("Posthumanist Performativity" 811). It "calls into question the givenness of the differential categories of 'human' and 'nonhuman'" (808). Barad's emphasis on the "intra-active" nature of being and knowing, rather than merely of representation, shows that reality is not constituted of pre-existing relations but of "things-in-phenomena" (*Meeting the Universe* 140). The actual capacity of matter is not that of a static entity, but is reconstituted in an ongoing way, as Barad's theory of "agential realism" explains. Based on this theory, the idea of matter "does not refer to an inherent, fixed property of abstract, independently existing objects; rather, *'matter' refers to phenomena in their ongoing materialization*" (Barad, *Meeting the Universe* 151). In an agential realist account, phenomena are "*the ontological inseparability/entanglement of intra-acting 'agencies'*" (139), which suggests, for example, a rethinking of climate change as an intra-active phenomenon. I argue that in relation to the performativity of climate change and intra-action among humans, sea creatures, animals, ice glaciers, storms, and Northern lights, as well as carbon emissions, bodies, elements, and chemicals, the composite body of Nuliajuk materialises, expresses itself, and catalyses unprecedented phenomena. Climate change itself is not a separate natural backdrop but functions as a performative apparatus that is entangled in the material-discursive history of the Arctic landscape and reconfigures that landscape's own agentic capacity. This perspective acknowledges that many other phenomena in the play's stories of ecological crisis, including the melting of glaciers, the threat of extinction for polar bears and other marine creatures, and the uncertain future of younger generations and their loss of identity and culture, cannot be taken for granted as pre-existing entities and independent relations. Rather, it is through intra-action with the changing climate, a range of material agencies, and human culture that each becomes a meaningful story and a form of "material narrativity" (Iovino and Oppermann, 8). Reading the play through this lens contributes to a materialist view of ongoing global change and a more perceptive approach to the earth's storied matter.

To consider how material narrativity is implicated in reading the play's derangement of scales, I focus on Barad's idea of the "intra-active topological dynamics" (*Meeting the Universe* 246). In rethinking the "spacetimematter" relations of climate change, to use Barad's terminology, geometrical concerns, which are built upon questions of size and shape, provide only one way of thinking about space; therefore, they are not sufficient for understanding the connectivity or topological concerns of different scales that are intra-actively made through one another (244). In the play, climate change crosses the manifold scales of individual bodies, such as those of Leanna's family, Jean, and the polar bears, the Inuit community, the Arctic region, and the globe, but these scales are not seen "as geometrically nested in accordance with some physical notion of size but rather are understood as being intra-actively produced through one another" (245). In this process, space and time are intra-actively created and reconfigured. In terms of temporal relations, for example, "the past is never left behind, never finished once and for all"; rather, the past and the future are enfolded by participating in "matter's iterative becoming" (Barad *Meeting the Universe* 181). In my view, Nuliajuk's body, which is affected by her own past history and human ecological damage, undergoes dynamic transformations, and its present participation in intra-activity with climate change interweaves past experiences with the possibilities of the future. Through the narrative history of Nuliajuk in this context of temporal expansion, the play expands, in Chaudhuri's words, the dramatic frame "temporally, pushing past cultural history to locate the human story in the deep time of the earth" ("Fifth Wall" n. pag.). Jean's entrapment in Nuliajuk's plant-like strands of hair suggests that while anthropogenic activities have reconfigured the environment throughout the Earth's history, human bodies are part of the world's continuing dynamism, through which, along with other bodies, they come to matter intra-actively. This performative space enacts a vital collaboration between the micro-scale individual human body and the larger macro-scale ecological mesh of imbricated material agencies, challenging the scope of human knowledge and subjectivity. In this way, the human is not the measure of all things, and what Barad refers to as posthumanism "is not captive to the distance scale of the human, but rather is attentive to the practices by which scale is produced" (*Meeting the Universe* 136). Rather than considering climate change a backdrop to our lives or a phenomenon in which we live, we need to reimagine how the personal and local scales of our bodies and daily lives are deeply entangled and co-existent with ecosystem and geological transformations. This perspective brings seemingly distant abstractions and statistics into intimacy with stories of breathing land, climate, the daughter bear's death, and Samuel's suicide.

Through its entangled engagement with the material world, the play raises the question of how we should take responsibility for the role that we play in the world's ongoing performativity. When Tulugaq finally goes after Jean and pulls him out of the sea, he tells Jean, who miraculously survived a near-death experience, that Nuliajuk is angry because of humanity's irresponsibility, and this causes storms that prevent Inuit from hunting and finding food. Jean needs to take responsibility, so he combs the sea goddess's hair, which calms the sea so that the sea animals are released. We can interpret Nuliajuk's anger and her demand to be combed as a call for ethical responsibility towards the environment and as a challenge to anthropic supremacy over nonhumans. We can also interpret this scene as the play's acknowledgement of our participation in the world's ongoing reconfigurations. As Barad argues, "[w]e are responsible for the world of which we are a part, not because it is an arbitrary construction of our choosing but because reality is sedimented out of particular practices that we have a role in shaping and through which we are shaped" (*Meeting the Universe* 390). By representing a world of intra-active becoming with which Jean and other human characters are in mutual relationship, the play calls for an ethics of responsibility rather than a detached objectivity about the changes of which we are a part. Jean's ethical commitment in combing the sea goddess's hair is a way of being sensitive and responsive to the environment and its possibilities for becoming. Further evidence of this ethical relation appears at the end of the play when, in his attempt to heal Veronica, Jean – who has also lost a child, in his case a stillborn child – offers her the polar bear sculpture to provide her with company. Jean undergoes a dramatic change in the course of the play from a scientist whose whole focus is on numbers and figures to a human who now understands how to care about other humans and living and nonliving entities. He takes a comb and, together with Leanna, begins combing Veronica's hair, which brings her to tears. At this moment, the sound of the song "Sila," by Inuit throat singer Tanya Tagaq, is heard while Nuliajuk's hair is loosening and releasing the Daughter bear and other creatures trapped therein. During the combing of Veronica's hair, which recalls Nuliajuk's mane of strands, a previously unacknowledged feeling for unseen spirits and forces is instead revered.

In response to this moment of acknowledgement and act of respect, the released animals "*gather in one of those unexpected formations only Nature knows how to create, and celebrate life with a water ballet*" (Bilodeau, *Sila* 99). Meanwhile, the Daughter bear swims up and then goes up into the sky and becomes a spirit (*anirniq*) while calling: "I'm here, Anaanaa. In the clouds, in the mountains, in the snow, in the ice" (105). In a deep intimacy between her spirit and that of the landscape, the cub drifts away and connects to the wider soul or Sila, just as Samuel's spirit is woven into it. Sila is a "common medium" for all the spirits in their immanent interrelation with each other (Abram 304). Whether a species is respected or threatened by the risk of extinction, in this shared space all the other humans, nonhumans, and the whole Sila are materially affected. The meaning of Sila becomes clear in the following conversation between Mama and Daughter:

MAMA: The only creatures who are lonely are the ones who forget about *sila*. . . . All life is breath. . . .

DAUGHTER: The sky is *sila*? *MAMA nods.*
The wind is *sila*? *MAMA nods.*
The land, the ice, the ocean? *MAMA nods.*

MAMA: And *sila* also moves in and out of our lungs. *MAMA breathes.*
See? That's *sila*. And with each breath, *sila* reminds us that we are never alone. Each and every one of us is connected to every other living creature. *The DAUGHTER breathes.* But *sila*'s gift is not ours to keep. We may use our breath while we roam the land but we must surrender it once we pass from the land. Creatures who are lonely are the ones who hold on to their breath as if it were theirs and theirs alone. (Bilodeau, *Sila* 43–4)

Our individual, but not autonomous or separate, breath is continuous with the natural environment, nonhuman animals, and objects. Each creature and material form is a manifestation of Sila that participates in this shared environment with its own unique perspective and awareness. According to David Abram, there is much affinity between "our experience of awareness" and "our felt experience of earth's atmosphere" (302). In the same way as our breath, Abram argues, "the awareness that stirs within each of us is continuous with the wider awareness that moves around us" and so "[e]ach organism partakes of this awareness from its own unique angle or situation within it, imbibing it through our nostrils or through the stomata in our leaves, altering its chemistry and quality within us before we breathe it back into the surrounding world" (303).[5] In this way, consciousness – one of the primary meanings of the word Sila – is not "the special possession of our species," but "a property of the breathing biosphere" (302) that articulates itself in interconnected stories of different forms of individual awareness or breath. As we are immersed in the earth's awareness and are conditioned by it, it is reconfigured by the changes we make with each breath. This perspective on Sila resonates with the concept of intra-action in terms of disturbing the subject-object dichotomy and revealing the world's iterative becoming and dynamism. The agency of Sila opens up a shift in our awareness of responsibility towards each breath we give to and take from the environment, recalling Abram's trenchant question: "[w]hat is climate change if not a consequence of failing to respect the elemental medium, air, in which we are immersed?" (301). Like the Covid-19 pandemic, the climate crisis threatens us all and then demands a united, collective, and immediate response. According to Achille Mbembe in his article "The Universal Right to Breathe" (2021), if the pandemic is "the spectacular expression of the planetary impasse in which humanity finds itself today, then it is a matter of no less than reconstructing a habitable Earth to give all of us the breath of life. We must reclaim the lungs of our world with a view to forging new ground. Humankind and biosphere are one" (S62).

The immanent mystery of Sila is performed once more, in the final scene, when Veronica delivers a spoken-word poem. The oral tradition of storytelling is central to Inuit culture, in part due to its ability to shape and transfer the exhaled breath and the fluid air, which are "the very medium of meaning" and "the implicit intermediary in all communication," and share them with those who listen (Abram 308). Veronica's work as a spoken word poet shows how the telling of stories allows every expressive actant – each of which has its own awareness and breath – to participate intra-actively in the surrounding environment and affect the story in the immediate moment of its narration. By sharing the stories of the sea, weather, and other human and nonhuman forces and by telling them out aloud, Veronica's performance, through its textures and rhythm, brings the human breath and awareness into an intimate conversation with other expressive actants and the expressive earth. This narration of exchange, consisting of manifold voices, composes the ongoing life of the Arctic landscape, despite the landscape's deviation from past sublimity due to the impact of climate change. While all the characters appear one by one and watch Veronica's performance, the play concludes with a new ecological awareness about human and nonhuman bodies as collective breath:

> *They all take a deep breath in . . .*
> *Projection (text):*
> Listen.
> I am right here.
> Inside of you.
> *. . . and let it out.*
> *Blackout.* (Bilodeau, *Sila* 105)

As this final moment of breathing out illustrates, the body, whether that of a human, polar bear, landscape, sea, ice, or the weather, is not figured as a self-determining entity with distinctive boundaries. Rather, in the iterative performativity of Sila, each body is entangled with the flow of other forces that are differentially produced and affected by the air that everyone and everything exhales. Seen in this light, the play's staging of inhaling and exhaling functions as an apparatus by providing a material space for the ongoing dynamics of intra-activity. This trans-corporeal environment is extended when each audience member also takes a deep breath, allowing us to grieve for the loss of human life, polar bear life, and the collective loss of our planet (Sandberg-Zakian iv–v). An understanding of collective grieving allows us to change our conception of extinction, as Veronica tries to do, and to accept extinction as a reality, in order to move forward and save what remains of our planet: "while yet we breathe it's never too late" (Bilodeau, *Sila* 104). Regarding this compelling feature of the play, the playwright expects audiences to "take center stage in the real world drama of climate" (Peach n. pag.). Given the necessity of this collaborative approach to living in the Anthropocene era, the play calls for alternative stories of becoming and expressing in an intimate interconnection with the breathing planet.

The ecodramaturgical features of the plays in this chapter, in addition to the works I have examined in the previous chapters, ask us to reconsider theatre's critical role in foregrounding silent voices and untold stories shaped by global ecological threats. These ecologically themed works invite us to rethink our bodily intra-action not only with other species but also with a broad spectrum of material forms and scales. Hence, the plays non-anthropocentrically perform hybrid forms of storied matter and material expression, reconfiguring the possibilities of what and where we are in the current era of the Anthropocene.

Notes

1 In response to why there is so little theatre on the climate emergency on the larger stages in the US, as "one of the major energy-gouging countries in the world," Bill Marx in his article "Theater Commentary: Why Are America's Stages Afraid of Dealing with the Climate Crisis?" argues that the reason might be partly "the stranglehold on our theater of

the domestic sphere" and the political "domination of large corporations" and their heavy investment in the fossil fuel industry (n. pag.).

2 For example, along with shooting, "dingoes are under sustained attack by pastoralists who mistakenly believe that the use of 1080 poison (sodium monofluoroacetate) will protect their vulnerable calves by diminishing the dingo population. In spite of scientific evidence, and in complete disregard of Indigenous people's views on the use of poison, the war against dingoes goes on" (Rose, *Wild Dog Dreaming* 56).

3 Considering the dictionary meaning of the term "sensitive" as "being '*quick to detect or respond to slight changes, signals or influences*,'" Bruno Latour states that this term "applies to Gaia *as well as* to the *anthropos*" ("Facing Gaia" 96).

4 In "From Landscape to Climatescape in Contemporary Dance-Theater" (2015), Amy Strahler Holzapfel suggests the term "climatescape" as "an outgrowth of landscape" (401). She argues that the very word "land," by suggesting a designated space and time on which theatre is conventionally founded, "feels oddly antiquated in comparison to 'climate,' which encompasses a sense of humans in relation to a larger, planetary spatiotemporal 'scape'" (401).

5 As David Abram argues, "Is consciousness really the special possession of our species? Or is it, rather, a property of the breathing biosphere. . . ? Perhaps the apparent 'interiority' we ascribe to the mind has less to do with a separate consciousness located somewhere inside me and another entirely separate and distinct consciousness that sits inside you and more to do with the intuition that we are both situated within it – a recognition that we are carnally immersed in an awareness that is not, properly speaking, ours, but is rather the earth's" (303).

References

Abram, David. "Afterword: *The Commonwealth of Breath*." *Material Ecocriticism*. Eds. Serenella Iovino and Serpil Oppermann. Bloomington: Indiana UP, 2014.301–14. Print.

Arons, Wendy and Theresa J. May. "Ecodramaturgy in/and Contemporary Women's Playwriting." *Contemporary Women Playwrights: Into the Twenty-First Century*. Eds. Penny Farfan and Lesley Ferris. New York: Palgrave Macmillan, 2013.181–96. Print.

Barad, Karen. *Meeting the Universe Halfway: Quantum Physics and the Entanglement of Matter and Meaning*. Durham, NC: Duke UP, 2007. Print.

———. "Posthumanist Performativity: Towards an Understanding of How Matter Comes to Matter." *Journal of Women in Culture and Society* 28.3 (2003): 801–31. Print.

Bilodeau, Chantal. "A Playwright's Journey to the Canadian Arctic." *World Policy Blog*. World Policy Institute, 19 Nov. 2014. Web. 01 Feb. 2016. <www.worldpolicy.org/blog/2014/11/19/playwrights-journey-canadian- arctic>.

———. "Digging for New Roots." *American Theatre*. Theatre Communications Group, 18 Feb. 2020. Web. 10 Sep. 2020. <www.americantheatre.org/2020/02/18/digging-for-new-roots/>.

———. "In Search of a New Aesthetic." Web blog post. *HowlRound*. HowlRound, 19 Apr. 2015. Web. 05 Mar. 2016. <http://howlround.com/in-search-of-a-new-aesthetic>.

———. "In 'Sila,' Perspectives on Climate Change Inspired by the Arctic." Interview by Linda Matchan. *Boston Globe*. Boston Globe Media Partners, 24 Apr. 2014. Web. 05 Mar. 2016. <www.bostonglobe.com/arts/theater-art/2014/04/24/sila-perspectives- climate-change-inspired-arctic/pj6GuCppO34GR6P2SosPDK/story.html>.

———. *Sila (The Arctic Cycle)*. Vancouver: Talonbooks, 2015. Print.

———. "Sila: An Artistic Contribution to the Problem of Climate Change." *The Center for Sustainable Practice in the Arts*. The CSPA, 16 June 2013. Web. 20 Mar. 2016 <www.sustainablepractice.org/tag/inuit-communities/>.

Cass, Loren R. "A Climate of Obstinacy: Symbolic Politics in Australian and Canadian Policy." *The Politics of Climate Change: Environmental Dynamics in International Affairs*. Ed. Paul G. Harris. New York: Routledge, 2009.11–28. Print.

Chaudhuri, Una. "The Fifth Wall: Climate Change Dramaturgy." Web blog post. *HowlRound*. HowlRound, 17 Apr. 2016. Web. 10 May 2016. <http://howlround.com/the-fifth- wall-climate-change-dramaturgy>.

———. "The Silence of the Polar Bears: Performing (Climate) Change in the Theater of Species." *Reading in Performance and Ecology*. Eds. Wendy Arons and Theresa J. May. New York: Palgrave Macmillan, 2012.45–57. Print.

———. *Staging Place: The Geography of Modern Drama*. Ann Arbor: The University of Michigan P, 1995. Print.

Chen, Szu-Feng. "Sila." *USITT-USA*. Prague Quadrennial, 2015. Web. 04 July 2016. <http://pq15.usitt.org/exhibition/submissions/scenery/sila-szu-feng-chen/>.

Citron, Cynthia. "A *CurtainUp* Los Angeles Review: *Song of Extinction*." *CurtainUp*. 2008. Web. 10 June 2015. <www.curtainup.com/songofextinctionla.html>.

Clark, Timothy. *Ecocriticism on the Edge: The Anthropocene as a Threshold Concept*. London: Bloomsbury Academic, 2015. Print.

———. "Phenomenology." *The Oxford Handbook of Ecocriticism*. Ed. Greg Garrard. Oxford, UK: Oxford UP, 2014.276–90. Print.

———. "Scale." *Telemorphosis: Theory in the Era of Climate Change, Volume 1*. Ed. Tom Cohen. Ann Arbor: MPublishing, Open Humanities P, 2012.148–66. Web. 14 July 2015. <http://dx.doi.org/10.3998/ohp.10539563.0001.001>.

———. *The Value of Ecocriticism*. Cambridge: Cambridge UP, 2019. Print.

Colebrook, Claire. *Death of the PostHuman: Essays on Extinction, Vol. 1*. Ann Arbor: MPublishing, Open Humanities P, 2014. Web. 11 Aug. 2015. <http://quod.lib.umich.edu/cgi/p/pod/dod-idx/death-of-the-posthuman-essays-on- extinction-volume-one.pdf?c=ohp;idno=12329362.0001.001>.

———. "Introduction: Extinction. Framing the End of the Species." *Extinction*. Ed. Claire Colebrook. Living Books About Life, Open Humanities P & Joint Information Systems Committee, 2012. Web. 16 Aug. 2015. <www.livingbooksaboutlife.org/books/Extinction/Introduction>.

Gray, Nelson and Sheila Rabillard. "Theatre in an Age of Eco-Crisis." *Canadian Theatre Review* 144 (Fall 2010): 3–4. <https://muse.jhu.edu/article/399617/pdf>.

Haraway, Donna J. *Staying with the Trouble. Making Kin in the Chthulucene*. Durham, NC: Duke UP, 2016. Print.

Heise, Ursula K. *Imagining Extinction: The Cultural Meanings of Endangered Species*. Chicago: University of Chicago P, 2016.

———. "Lost Dogs, Last Birds, and Listed Species: Cultures of Extinction." *Configurations* 18.1–2 (Winter 2010): 49–72. Print.

———. *Sense of Place and Sense of Planet: The Environmental Imagination of the Global*. New York: Oxford UP, 2008. Print.

Holzapfel, Amy Strahler. "From Landscape to Climatescape in Contemporary Dance-Theater: Meredith Monk, The Wooster Group, and The TEAM." *The Oxford Handbook of Dance and Theater*. Ed. Nadine George-Graves. Oxford: Oxford UP, 2015.401–28. Print.

Huntington, Henry, et al. "An Introduction to the Arctic Climate Impact Assessment." *Arctic Climate Impact Assessment*. Eds. Carolyn Symon Lelani Arris and Bill Heal. Cambridge: Cambridge UP, 2005.1–19. Print.

Iovino, Serenella and Serpil Oppermann. "Introduction: Stories Come to Matter." *Material Ecocriticism*. Eds. Serenella Iovino and Serpil Oppermann. Bloomington: Indiana UP, 2014. Print.

Latour, Bruno. "An Attempt at a 'Compositionist Manifesto'." *New Literary History* 41 (2010): 471–90. Print.

———. "Facing Gaia: Six Lectures on the Political Theology of Nature." Gifford Lectures on Natural Religion. Edinburgh. 18–28 Feb. 2013. Web. 16 Sept. 2015. <https://macaulay.cuny.edu/eportfolios/wakefield15/files/2015/01/LATOUR-GIFFORD-SIX-LECTURES_1.pdf>.

Lewis, E.M. "Pictures of Song of Extinction from the EcoDrama Festival." Web blog post. *Prayers to Broken Stone. . ..* E.M. Lewis, 9 June 2009. Web. 24 Dec. 2016. <http://prayerstobrokenstone.blogspot.com.au/2009/06/pictures-of-song-of- extinction-from.html>.

———. *Song of Extinction.* New York: Samuel French, 2010. Print.

Marx, Bill. "Theater Commentary: Why Are America's Stages Afraid of Dealing with the Climate Crisis?" *The Arts Fuse.* The Arts Fuse, 20 May 2020. Web. 10 Sep. 2021. <https://artsfuse.org/201969/theater-commentary-why-are-americas-stages-afraid-of-dealing-with-the-climate-crisis/>.

May, Theresa J. "Radical Empathy, Embodied Pedagogy, and Climate Change Theatre." Web blog post. *HowlRound.* HowlRound, 20 Apr. 2016. Web. 10 May 2016. <http://howlround.com/radical-empathy-embodied-pedagogy-and-climate-change-theatre>.

Mbembe, Achille. "The Universal Right to Breathe." *Critical Inquiry* 47. S2 (Winter 2021): S58–S62. <www.journals.uchicago.edu/doi/full/10.1086/711437>.

McGraw, Forrest. "(Eco)Genocide in Cambodia." Web blog post. *Political Ecologies.* WordPress, Dec. 2012. Web. 25 June 2015. <https://forrestmcgraw.wordpress.com/2014/07/07/red-to-blue-to-green-ecogenocide- in-cambodia>.

Merkur, Daniel. "Breath-Soul and Wind Owner: The Many and the One in Inuit Religion." *American Indian Quarterly* 7.3 (Summer 1983): 23–39. Print.

Nicholson, Amy. "Song of Extinction." *Los Angeles Weekly.* LA Weekly, 13 Nov. 2008. Web. 20 July 2016. <www.laweekly.com/calendar/song-of-extinction- 2156964>.

Nixon, Rob. "The Anthropocene: The Promise and Pitfalls of an Epochal Idea." *Edge Effects.* Nelson Institute, 6 Nov. 2014. Web. 30 June 2016. <http://edgeeffects.net/anthropocene-promise-and-pitfalls/>.

O'gorman, Emily, et al. "Teaching the Environmental Humanities: International Perspectives and Practices." *Environmental Humanities* 11.2 (Nov. 2019): 427–60. <https://read.dukeupress.edu/environmental-humanities/article/11/2/427/140787/Teaching-the-Environmental-HumanitiesInternational>.

Peach, Sara. "Tales for a Changing Arctic." *Yale Climate Connections.* The Yale Center for Environmental Communication, 05 Aug. 2016. Web. 15 Jan. 2017. <www.yaleclimateconnections.org/2016/08/tales-for-a-changing-arctic/>.

Picq, Manuela. "Bolivian President Evo Morales's Ecocide Is a Genocide." *Truthout.* Truthout, 26 Sep. 2019. Web. 10 May 2020. <https://truthout.org/articles/bolivian-president-evo-morales-ecocide-is-a-genocide/>.

Richards, Kimberly and Heather Davis-Fisch. "Editorial: Extractivism and Performance." *Canadian Theatre Review* 182 (Spring 2020): 5–8. <https://ctr.utpjournals.press/doi/full/10.3138/ctr.182.001>.

Rose, Deborah Bird, et al. "Introduction: Telling Extinction Stories." *Extinction Studies: Stories of Time, Death, and Generations.* Eds. Deborah Bird Rose et al. New York: Columbia UP, 2017.1–17. Print.

Rose, Deborah Bird. "Monk Seals at the Edge: Blessings in a Time of Peril." *Extinction Studies: Stories of Time, Death, and Generations.* Eds. Deborah Bird Rose et al. New York: Columbia UP, 2017.117–46. Print.

———. “Slowly ~ Writing into the Anthropocene.” *TEXT* 20 (Oct. 2013): 1–14. <www.academia.edu/5081786/Slowly_Writing_into_the_Anthropocene>.
———. *Wild Dog Dreaming: Love and Extinction*. Charlottesville and London: University of Virginia P, 2011. Print.
Sandberg-Zakian, Megan. “Introduction.” *Sila (The Arctic Cycle)*. Ed. Chantal Bilodeau. Vancouver: Talonbooks, 2015.i–v. Print.
Slunge, Daniel and Susanne von Walter. “Environment and Climate Change in Bolivia: Challenges for Opportunities and Development.” *Sida's Helpdesk for Environment and Climate Change* (May 2013): 1–36. Print.
Soper, Ella and Nicholas Bradley, eds. *Greening the Maple: Canadian Ecocriticism in Context*. Calgary: University of Calgary P, 2013.
Stern, Pamela R. *Daily Life of the Inuit*. Santa Barbara: Greenwood, 2010. Print.
van Dooren, Thom. “A Howling Lamentation.” Web blog post. *Thom van Dooren*. 24 July 2019. Web. 10 Oct. 2021. <www.thomvandooren.org/2019/07/24/a-howling-lamentation/>.
———. “Vulture Stories: Narrative and Conservation.” *A Manifesto for Living in the Anthropocene*. Eds. Katherine Gibson, Ruth Fincher and Deborah Bird Rose. New York: Punctum, 2015.57–62. Print.
Yeo, Sophie. “Chantal Bilodeau Brings Climate Change to the Theater.” *Pacific Standard*. Grist, 09 July 2018. Web. 02 Oct. 2019. <https://psmag.com/environment/chantal-bilodeau-bringing-climate-change-to-the-theater>.

5 Setting the stage for possibilities of collaborative survival

Jordan Hall's *A Brief History of Human Extinction* and Miwa Matreyek's *This World Made Itself* and *Infinitely Yours*

This chapter analyses *A Brief History of Human Extinction* (2018) by Canadian playwright Jordan Hall and *This World Made Itself* (2013) and *Infinitely Yours* (2020) by Miwa Matreyek, a multimedia performance artist from Los Angeles. These provocative works of theatre and performance engage thematically, performatively, and experientially with the idea of the Anthropocene, challenging audiences to think differently about the history of the earth, human relationships with nonhumans and matter, and environmental crises. This chapter, like Chapter 4, argues that the Anthropocene crisis demands representational changes in theatre. Hall's play and Matreyek's performance art ecocritically investigate the representational logic of depicting the end of human life on an uninhabitable planet and the problem of continuing to live in this changing world. These theatrical works embrace different representational strategies to stage the collaboration between theatre, art, and the Anthropocene. While Hall's play – created in collaboration with Mind of a Snail puppet theatre – employs mixed media, science fiction, and speculative art in theatre, Matreyek combines the ecodramaturgical practices of collage animation, dance, musical elements, installation art, film, theatrical performance, and eco-activism. These works prompt the audience and participants to see and feel perspectives beyond their own by engaging with issues of ecological complexity that are enacted through the performances. While Hall explores the idea that humanity is caught in the midst of vast temporal and spatial scales of planetary mass extinction and that this necessitates an interrogation of anthropocentric hierarchy, Matreyek's interactive performances constitute scientific, ecological, emotional, visual, and embodied meditations. Her integration of her own shadow into animated images of earth history, evolution, and environmental degradation prompts viewers to reflect on the entanglement of their bodies with a damaged and damaging ecology. More broadly, Matreyek's multimedia works make a fittingly dramatic way to close this book. Her works seem to juxtapose all the material and Anthropocenic images that we have seen in the previous plays in a collage, by delineating a journey through the earth's history,

DOI: 10.4324/9781003048749-5

sketching anthropogenic impact on the earth, weaving material narratives of intra-action among humans and nonhumans, and performing a "geostorical" form of narration "inside which all the former props and passive agents have become active without, for that, being part of a giant plot written by some overseeing entity" (Latour, "Facing Gaia" 3). Through a geostorical lens, the performances' dynamic process of making-with envisions a compelling map of how we humans might reach a different way of living on, in, and with the material world.

1 The trouble of living and dying together on a wounded planet in *A Brief History of Human Extinction*

A unique collaboration between Hall and the multimedia theatre artists of Mind of a Snail Puppet Company, *A Brief History of Human Extinction* (originally titled *Rate of Loss*) premiered at Vancouver's Upintheair Theatre in October 2018. In an email conversation with Hall, she indicates that this play and much of her work have been influenced by the dual reality of Canada as a nation:

> We are often defined by our natural environment – our identity as a nation is synonymous with a kind of vast wilderness, stark and beautiful. . . . But we are also a huge driver of climate change – our per capita emissions are huge, and the Alberta oilsands have been described as a 'climate bomb.' I don't identify strongly with my nation, but to be part of this culture is to live in that dichotomy: We lionize the natural world, and we are destroying it. I think a lot of *Brief History* was about exploring the grief of that. And exposing the false consciousness – the artificial separation we try to make between ourselves and the natural world – that makes the exploitation possible.
>
> (Hall, "Notes")

Considering this national and regional configuration, the play paints a portrait of a world affected by vast, interconnected anthropogenic ecological crises. Set 160 years in the future, the play shows that climate change, species extinction, and an invasive fungal plague have rendered the surface of the planet uninhabitable. Ever Pickwas-Holtz and Adam, the last remnants of the human race, are trapped inside a decimated dome or, as Ever says, "a sealed biosphere" (Hall, "Human Extinction") that shields them from the poisonous outside world. As a way of dealing with species loss on planet Earth and carrying out the mission that they have been assigned, Ever and Adam prepare for the launch of a rocket called the Ark, an embryo ship or capsule that will carry the genetic legacy of all sorts of earthly creatures – including *Homo sapiens* – to a new home on a distant planet (Kepler 186F). Their only companion is Ommie, the last living sea otter, swimming around in its tank. In the play's 2018 production by Upintheair Theatre, Stephansie Elgersma skilfully designed the otter as a life-size puppet manipulated by two or three actors (see fig. 5.1). Adam, a genetically engineered maintenance worker and Ever's childhood friend, maintains the biosphere like a farm or a zoo, cultivating his garden spaces and performing the menial labour of sanitising and maintaining the domes. When we first meet Ever, the "Project Lead" of the Next Earth Foundation's Human Expansion Project, she is recording video messages for the potential future colonists of Kepler. For Ever, this is "the final message from humanity on Earth to its galactic future. . . . It's all they're going to have – on some alien world, hundreds of years from now. This story of who we were" (Hall, "Human Extinction"). In her first monologue, Ever reminds the audience of a long history of anthropogenic activities and a natural world irrevocably altered by their ecological impacts, including "Sea-Level Rise and Ecosystem Collapse and Mass Extinction," and the "new diseases" that come with them (Hall, "Human Extinction"). However, she believes in the unlimited potential of the human species in spite of such mistakes, and insists on launching the Ark, which contains genetic material, viable DNA from humans and other creatures. But while Ever's message for future generations may be expected to shape our new beginning, as the launch of the Ark gets closer a deep uncertainty about a possible future and new life for humanity is growing within her. Adam, meanwhile, is

Figure 5.1 Lisa C. Ravensbergen as Ever, Daniel Martin as Adam, and otter. *A Brief History of Human Extinction*. Upintheair Theatre, 2018. Directed by Tamara McCarthy. Projection design by Mind of a Snail. Photo by Matt Reznik.

beginning to realise that launching the Ark will use up the remainder of the fuel that powers the domes, exposing himself and Ever to the lethal environment outside the biosphere. Adam wants to use the rocket's limited resources and supplies so that he and Ever can survive longer on earth. He is also incredibly frustrated by Ever's seemingly wasteful compulsion to keep Ommie, the last living sea otter, alive a kind of companion. Adam objects to the presence of the otter as the third inhabitant of their dome and argues that following Ever's plan and launching the Ark is essentially suicidal. These tensions and uncertainties are exacerbated by the sudden and terrifying presence of something mysterious outside the facility, threatening to come in. As a fungal plague has eradicated almost all life on earth, the play reminds us of this dark reality: "Extinction. It's the only way the story ends. You either die or live long enough to evolve into something that isn't you anymore" (Hall, "Human Extinction").

As a follow-up to Hall's award-winning plays *Kayak* (2011) and *How to Survive an Apocalypse* (2016), *A Brief History of Human Extinction* explicitly provides an innovative theatrical approach to a global ecological crisis that crosses species boundaries. For the playwright, who has also contributed to three cycles of the Climate Change Theatre Action (CCTA), "many of the strategies we use to create narratives around the climate crisis are still drawn from ancient and classical models – which often feel entirely inadequate when addressing the scale and nature . . . of the issue" (Hall, "Eco-Theatre" n. pag.). Hall asks, "How do you write about a radical new moment, the stakes of which are nothing less than the future of your species (and many others) when the form you practice seems to resist the story you want to tell?" ("Eco-Theatre" n. pag.). To dramatise a different, more effective framework for narrating a story of planetary doom, Hall created this play in collaboration with the multimedia theatre artists of Mind of a Snail. As a result of this collaboration, skilled forms of hand puppetry and shadow puppetry, live action, video streaming and projection, cartoons, and narrative address are combined in the production. More significantly, this jointly created production "echoes the interplay between simple and complex life forms in its interplay between two- and three-dimensional realities" (Oliver n. pag.), while communicating about the earth's future.

In the play, Hall and Mind of a Snail develop a new form of eco-theatre that acknowledges the seriousness of earth's ecological challenges over the next 160 years and integrates science fiction into performance art. This fusion offers a different interpretation of the Anthropocene concept and a new strategy in eco-theatre by foregrounding the changing patterns of spatial and temporal scales, with a sense of the scale of mass extinction. Chloé Ziner, the founder of Mind of a Snail Puppet Company, thinks that the play's employment of science fiction is "a way of addressing contemporary topics with immediacy" (Derdeyn n. pag.). To foreground this urgency and immediacy, I organise my discussion of the play around the following questions: What strategy does the science fiction play develop to tell stories about long time spans in the Anthropocene? How does the play represent human and nonhuman forms of agency and their ongoing entanglements in the age of mass extinction? This is where the consideration of comedy as a genre belongs. The play uses comedy – particularly via the figures of Ever and the fungus – to undercut the grandiose narratives of sci-fi, which are so often about human exceptionalism. The interplay of this storytelling approach with Mind of a Snail's props, sets, and movement pieces – which would have been huge, beautiful designs and special effects in a sci-fi film, but are toy-like, clearly handmade objects here – echoes the comedic undermining of grand human ambition (Hall, "Notes"). While Lewis, in *Song of Extinction*, uses the framework of elegy to dramatise losses due to Anthropocenic mass extinction, Hall's play relies upon the imaginative and rhetorical tools of science fiction as a "form of contemporary epic storytelling" (Heise, *Imagining Extinction* 215) to scale up our imagination of human-nonhuman relationships in light of extinction. Hall is also drawn to Ursula Heise's sense of comedy as a form that "focus[es] on the loss and regaining of equilibrium after upheaval," helping us "to start telling new stories about extinction and climate change" (Hall, "Eco-Theatre" n. pag.). Hall indicates that the play is "a sci-fi comic thriller about the last adventure that anyone will ever have" (Derdeyn n. pag.). The play seems like it might arguably imagine both apocalyptic and salvific futures – after all, it represents extinction and the death of two humans and Ommie on earth, and a rocket that launches for another planet carrying earthly DNA. However, Hall ironises and undercuts those moves by provoking and challenging the audience to think critically about the possibility of persisting with the trouble, being truly present, and responding to disastrous events. As Hall adds, the play "has reality firmly embedded in it," leaving audiences "with the clear feeling that this is actually really happening right now, that this vastness of destruction is real and that the best thing we can do is find ways to adapt" while being "able to find ways to play in it too " (Derdeyn n. pag.). The play criticises longings for an idealised past or a magical future and despair about the end of all things, instead calling for us to be in the present and develop insights into and respond to this challenging situation through ongoing entanglements with other humans and nonhumans.

Science fiction theatre and the changing patterns of spatial-temporal scales in the Anthropocene

The play challenges us to reconsider what "the human" means on a changing planet in a time of mass extinction. As the play begins, we hear the musical composition "Next Earth," played by Nancy Tam, which is followed by Ever's monologue about the sterility of life on planet Earth, described as "a fungal wasteland, desolate and rotting" (Hall, "Human Extinction"). Ever emphasises the exceptionality and unique rights of *Homo sapiens*, who she sees as worth saving through the potential colonisation of a new planet, and who she hopes are capable of learning from the irreversible and disastrous damage that they have inflicted on this planet. Further, Ever declares that future humans are the exception to the dark reality of species loss and extinction: "But not you. That's not your story. You are the genetic continuation of the intelligent primate species *Homo sapiens sapiens*. You exist in an unbroken chain of Life from that first tiny spark, through all the beings who have been our ancestors, through me, to you" (Hall, "Human Extinction"). The question of human exceptionality and autonomy is at the critical heart of this apocalyptic sci-fi play.

While Ever acknowledges that many – thousands? millions? – of species have been affected by mass extinction, she has placed her faith in human ingenuity: technology and progress as a way to move towards the next stage of evolution. For example, the play's allusion to Stanley Kubrick's *2001: A Space Odyssey* emerges from Ever's obsession with human grandeur and intelligence:

ADAM: I like the part where the robot goes nuts. And you like it. . . .
EVER: It's a metaphor – for evolution and progress –
ADAM: The magic rock?
EVER: Well. Yes. . . . The evolution of intelligence beyond us. That's why the computer goes nuts. It's another competitor for the intelligence niche. We outcompete it and evolve into the Star Child and – (*Off the confusion on his face:*) Or – or it's like Ommie.
ADAM: Ommie is a magic rock?
EVER: She's our potential. We hunted so many species to extinction. We didn't care enough to save so many others. But her? We passed laws and ran captive breeding programs for her ancestors and – she's here. Hundreds of years of conservation efforts distilled into one last otter. She's what we're capable of when we're our best selves. (Hall, "Human Extinction")

This evolution into the "Star Child," or new life, that Ever imagines is borne out of a deep faith in human intelligence and exceptionalism, as demonstrated by actions like the conservation of endangered species such as the sea otter – hence Ever's attachment to Ommie and what she represents. This conservation effort is also linked to the rational progress and evolution of humanity through science and technology. However, as the play reaches its end, Ever and Adam realise that technological discovery cannot stop the intrusion of nonhuman forces into human life or help the characters save what they consider crucial to the human experience. Invoking the idea of the extinction of humankind, the play raises an ecological question about what would happen to our planet, altered profoundly by human activities, if humans themselves disappeared. Seen from this perspective, Heise argues that "Earth itself appears as a terraformed planet – a planet reshaped to suit the ideas and needs of humans and yet out of their ultimate control" (*Imagining Extinction* 216). In other words, the world as we know it has ended and we live on an alien planet, which acts in an unpredictable and uncertain way. The lighting and set design created by Jerguš Opršal, including "cracked green-tinted walls and a tank for the otter," provides the simple but effective atmosphere of a barren landscape (Gartlan-Close n. pag.). Further, Tam's music and sound design intensifies the sense of imminent doom and dark horror resulting from a transformed planet.

To question the idea of human transcendence and suggest instead that large-scale changes in the global environment will transgress human boundaries, Hall sets the human characters the challenge of dealing with the entanglements of technology, human progress, biodiversity loss, and shifting geological time scales. An example of such challenges is the conflict between Adam and Ever over launching the Ark or attempting to build a community on Earth. This conflict becomes more intense as they are attacked by fungally infected crows, who breach one of their enclosures and spread the invasive fungus amongst Adam's carefully selected farm animals and crops. The creeping shadow of the strange presence outside now comes inside, contaminating the whole orchard enclosure:

> *There's a sound from the enclosure window. Something like a hand pressed against the window. Adam looks, and there, waving against the window, just like the branches of his trees, is a hand-shaped spore tendril. Adam peers closer. The spores open their eyes. Adam reels back in horror and incinerates it all.*
>
> (Hall, "Human Extinction")

The horrible scene in which the orchard is burned is followed by an argumentative dialogue: While Adam argues that they need to use the launch packet and its supplies, and get rid of anything non-essential, including Ommie, to stay alive on earth, Ever, who does not care about the orchard, insists on getting species to the planet Kepler, where the new human colonists will be able to replicate the livable ecology and environment of the past Earth on a whole new world:

ADAM: We can't afford the drain. To run her tank. To feed her her own goddamn weight in protein every five days. . . .

EVER: I'm not killing the last otter on earth because you got drunk and hallucinated religion.

ADAM: It wouldn't be enough anyway. The reserves are all gone. We need more fuel so we can defend ourselves. We need more fuel to rebuild –

EVER: No. We don't touch the launch packet.

ADAM: Why not? . . . You're not even sure you should send them. I hear you talking, you know – to the goddamn Otter. The goddamn Otter that you talk to more than you talk to me. You're scared they might be monsters. You're scared they're just going to land on Kepler and kill everything just like we did before –

EVER: Kepler is the only future I can give them –

ADAM: What about us? What about our future?

EVER: God. Adam, We have no future! We are terminal animals. We are specimens living in a cage because we cannot survive in our environment. We are unfit. We made a world in which we are unfit.

ADAM: But we have the domes. With enough time and fuel we could –

EVER: What? Repair them? Build new ones? Start ourselves a little hermetically sealed town? Even if we rebuilt the orchard dome and used the Ark to spin ourselves up some kids and grandkids – it would be at the expense of all those lives on the Ark. Their chance at a whole new world. We can't be that selfish and be worth saving at all. I'm going to get them to Kepler. I'm going to try. (Hall, "Human Extinction")

This conversation raises a question about the extent to which humans are able to shape their own futures and the extent to which their futures are determined by the anthropogenic global ecological problems of which they are a part. Adam's concern about his future and Ever's on earth and his attempt to convince Ever to stay and rebuild their current habitat contrasts with Ever's concern about the future of all those lives on the Ark as well as her increasing skepticism about humanity's future, including their own. Their conflict mirrors the complexity of the vast spatial, temporal, and material scales of the global extinction crisis. In this way, to use Heise's words, the play embodies the "stories of large-scale change and forces influenced by and yet not controlled by humans" ("Time Scales" 301). In such stories, Heise argues,

> [t]he larger-than-life hero or single protagonist may decrease in importance, since epic-style narratives over the last century have tended to shift the major narrative actants from individual human characters to collective and sometimes nonhuman actors. Individual humans as minor characters will no doubt continue to abound – but they may end up being just that, minor characters.
>
> (301)

Drawing upon the genre of science fiction, the play scales up the imagination of the human through engagement with the fate of planet Earth as a whole, and projects the time scales of both the deep past and distant futures on the Earth itself. The play thus historicises the human present from the perspective of a far future and represents individual human stories through their connections with massive evolutionary time scales and staging. Seen in this light, Adam and Ever are reduced to the status of minor characters against the vast spatial and temporal span of ecological change while the play enacts ecological awareness, prompting the audience to be skeptical about human agency over the long term. As Heise argues, such protracted temporal patterns "raise the question to what extent the individual characters truly make decisions of their own or are subject to a pattern that they may not be aware of" (288). In the play, the vastness of spatial-temporal scenarios highlights human vulnerability to the Earth's geological evolution, creating a sense of unpredictability.

Considering this science fiction play in relation to the earth's deep-time history deepens the challenge of redefining what "the human" means. In order to examine this challenge, I focus in more detail on the conflict between Ever and Adam. In response to Ever's willingness to sacrifice his present existence in favour of the implausible hope of creating a utopian ecosystem of her own design, Adam allows Ommie to escape her tank. He thinks that by allowing Ommie to die in the lethal environment outside the dome, he and Ever will have more resources to keep themselves alive longer. The following conversation clarifies the contradictions that the human characters grapple with:

ADAM: God! Ever! Everything alive is selfish! You think those kids on your Ark aren't going to be selfish? You think they're going to listen to your stupid videos and think: "Oh, some lady in the past who was too dumb to even keep herself alive says we're bad, so let's just ignore like a billion years of evolution?" No. They're going to look at all those samples you sent with them, and they're going to know what I know. That we have Dominion. Life exists to serve us. To be meat and eggs and milk to us. That's why you're not just sending a bunch of human samples to Kepler. They exist to keep us alive.

EVER: No. They can learn from our mistakes. They can be better –

ADAM: What? Like you're better? You think the exact same evolution didn't make you? You're worse than all of them. Sitting up here, looking down on me because I love you. Because I care about you more than about frozen samples or some goddamn otter. Because I want you to live? (*Beat.*) You'll be lucky if they're only as selfish as me. At least then they won't be hypocrites. . . .

EVER: You're right. They're us. All those little possibilities. . . . Why pretend about it? That's what I should just tell them. I should just – (*She scrambles over to the camera and pulls up her recording. Intro Ping:*) Hey Kids! Here are the facts: You are the descendants of mean, murdery little monkeys, who spent most of their time exploiting and killing other creatures and each other for good measure. . . . we could cooperate – but we mostly just cooperated to get better at the killing and the exploiting. We built our civilizations on huge unsustainable mass harvests. . . . We invented slavery, and torture, and freaking genocide – . . . and used a bunch of fairytale religions about how goddamn special we were to justify all the shitty things we did –

ADAM: C'mon, stop it –

EVER: And it was all a great warm-up for our big number: The Mass Extinction of pretty much ALL the life that just happened to be sharing a planet with us! In all honesty, kiddies, probably the best thing that could happen for the future of life on this planet and anywhere else, is our complete and total extinction! (Hall, "Human Extinction")

Having pondered Ever's description of the era of anthropogenic species extinction from the perspective of deep time, the audience concludes that natural history is not singular or separate from human history. Also, thinking about humanity in relation to the scale of geological history leads the play to consider a series of global ecological changes, such as mass extinction. For example, Adam's emphasis on the idea that humans have dominion, that they can remake their paradise on earth, that life exists to serve them, and that nonhuman species are simply resources for humans is juxtaposed with Ever's concern about mass extinction and the end of the world (or at least the end of the human world), which result from human intervention in the global environment. However, thinking about human agency in relation to the spatial and temporal framework of geology challenges the centrality and prominence of that agency. Ever's reference to "our complete and total extinction" diminishes humans' grandeur, begging the question of what a world without humans would be like. As Claire Colebrook argues, "What happens if one thinks of the vision of no one, of the human world without humans that is still there to be seen?" (*Death of the PostHuman* 28). Colebrook explains the question by placing humans in the long time span of the Anthropocene:

> The positing of an anthropocene era . . . deploys the idea of human imaging – the way we have already read an inhuman past in the earth's layers – but does this by imagining a world in which humans will be extinct. The Anthropocene thought experiment also alters the modality of geological reading, not just to refer to the past as it is for us, but also to *our* present as it will be without us.
>
> (*Death of the PostHuman* 28)

This perspective reconceptualises the human in the Anthropocene by putting an end to human exceptionalism and individualism and foregrounding the agency of the global species extinction crisis in which we are involved. Further, the play's biblical references can be linked to the concept of the Anthropocene. The fall of Adam and Eve and their sins towards each other and the environment might be regarded as one of the origins of a fatal flaw – the anthropocentrism that leads us to see ourselves as superior to the rest of creation. Our anthropocentric tendencies make us capable of destroying our planet to the extent that it threatens species including our own and the whole ecosystem, radically inverting the history of geology and inaugurating the Anthropocene.

The play's invitation to look at our present through what Hall calls "interweaving time frames" (Derdeyn n. pag.) and unfolding, entangled human-nonhuman futures evokes a different perspective on narratives about the Anthropocene idea. In her article "We Have Always Been Post-Anthropocene: The Anthropocene Counterfactual" (2017), Colebrook addresses the necessity of such new forms of narrative by comparing the Anthropocene with Darwinian evolution:

> Rather than Darwin's timeline of life's grandeur, with a (random but fortuitous) proliferation of difference and complexity, we might have to confront a sudden event of geological impact within an intensively human timeline. Rather than one more event within time, not only does the Anthropocene – as did Darwinian evolution – require us to shift our scale of narration away from human generations and history to species' emergence and deep time; it also raises the problem of intersecting scales, combining the human time of historical periods (late capitalism, industrialism, nuclear power) with a geological time of the planet.
>
> (1)

As a speculative performance, the play dramatises the problem of intersecting timescales that Colebrook raises through a glimpse of the transformed planet's future that is, indeed, a projection of the present. The play's intersecting timescales also reveal how the Anthropocene extinction prompts us to not only ponder the history of human society, but also imagine the legacy we will leave thousands of years into the future. Heise argues that the Anthropocene itself might be understood as geological speculative fiction because it "focuses on the reality of a terraformed planet that the genre has held out as a vision for the future of other planets, but which has already arrived in the present on our own planet" (*Imagining Extinction* 219–20).[1] From this perspective, the play exemplifies "the idea of a planet terraformed by humans in such a way that the traces of the process will be perceptible in the geological strata to a putative far-future observer" (Heise, *Imagining Extinction* 217–18). As a way of dealing with the global species extinction crisis, the play explores "de-extinction" and "rewilding," which materially perform "broader scenarios of terraforming, the shaping of global ecosystems at the hands of humans" (212). In order to send the embryonic human colonists and their host of support species to Kepler, Ever tests and collects genetic samples of many extinct species representing the biological legacy of planet Earth. She lists these samples and after each species, she gives the cause of its extinction: "*Pseudacris maculata*. Boreal Chorus Frog. Chytrid Fungus and Seasonal Shift. Samples viable. *Ammospiza maritima*. Seaside Sparrow. Destruction of Habitat and Pollution. Samples viable. *Engraulis mordax*. Anchovies. Overfishing" (Hall, "Human Extinction"). And in another part of the play, Ever lists other cases: "*Eubalaena australis*. Southern Right Whale. Uh – Overfishing and Ocean Pollution. Samples Viable. *Canis Rufus*. Red Wolf. That was – they were – Exterminated by Ranchers. Samples Viable. *Homo sapiens sapiens*. D3 Fungus. . . . Self-inflicted" (Hall, "Human Extinction"). While Ever's experiment is improbable, conceptualising it in terms of the prospects of de-extinction and rewilding shows how Ever aims to create biospheres composed of new habitats for both humans and nonhumans in the present and future, rather than to restore past ecosystems. This effort reminds the audience of the seriousness of earth's ecological crisis over the next few hundred years as well as the impossibility of any return to the ecological past (see fig. 5.2).

In light of the terraforming natures of science fiction, the play responds to the following key questions that arise in the current era of human-driven mass extinction: "[w]here do we begin on such final endings? And in what ways might we actually convey the meaning of such endings?" (Jones et al. 390). This section has answered these questions in part by linking the play's speculative art with the spatiotemporality of mass extinction and the problem of deep-time history implicit in the Anthropocene idea. The play invokes questions about the place of human beings in the world, turning individual characters into minor characters by placing them on a spectrum of scales and shifting agency away from them to nonhuman actors. As Heise points out, "[t]he clashing scales of geological time, of the microbes and bacteria that make up part of human bodies, and of the chemical flows of nutrients and toxins through organisms, . . . all lead to different reconceptualisation of human individuality and collectivity" ("Time Scales" 277). As I respond more fully to the questions above in the following section, I examine how the various scales of human and nonhuman, individual and global are materially interconnected to provide a broader perspective on the age of mass extinction.

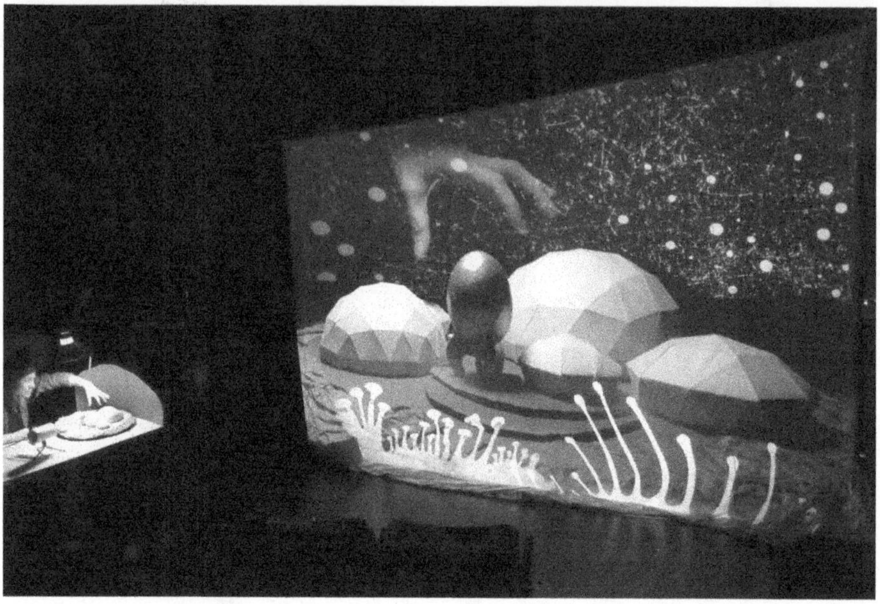

Figure 5.2 A Brief History of Human Extinction. Upintheair Theatre, 2018. Directed by Tamara McCarthy. Projection design by Mind of a Snail. Photo by Itai Erdal.

The stage of "making oddkin" through the collaborative mapping of extinction

As the play moves on, its spatial and temporal flexibility opens up possibilities of thinking, narrating, and performing outside the human perspective, asking us to reconsider human exceptionality through confrontations with the nonhuman other in the form of the "Stranger." In different parts of the play, Adam and Ever sense an unseen presence: "*Beware the Stranger . . . And the Stranger shall appear at your gates . . . For he will come as a storm in the wasteland . . . A storm of eyes and wings and fire*" (Hall, "Human Extinction"). The "Stranger" entities are performed by puppeteers at the edge of stage and appear in the toxic, deadly form of fungal tendrils that infect and threaten all life forms, including the orchard enclosure, the Ark, the whole facility, and in particular Ommie[2] (see fig. 5.3). "*Ommie emerges out onto the vast fungal wasteland, and the figure of the fungal otter coalesces before her, blinking its many eyes, just as the hand in the window of the orchard dome did. . . . Ommie and the fungus reach towards each other, closer, and closer*" (Hall, "Human Extinction"). This hybrid creature – the otter entangled with the fungal bodies – re-emerges from the intake pipe, from which they continue their slow interpenetration of the stage space. This kind of entanglement unfolds the possibility of "sympoietic" relationships within

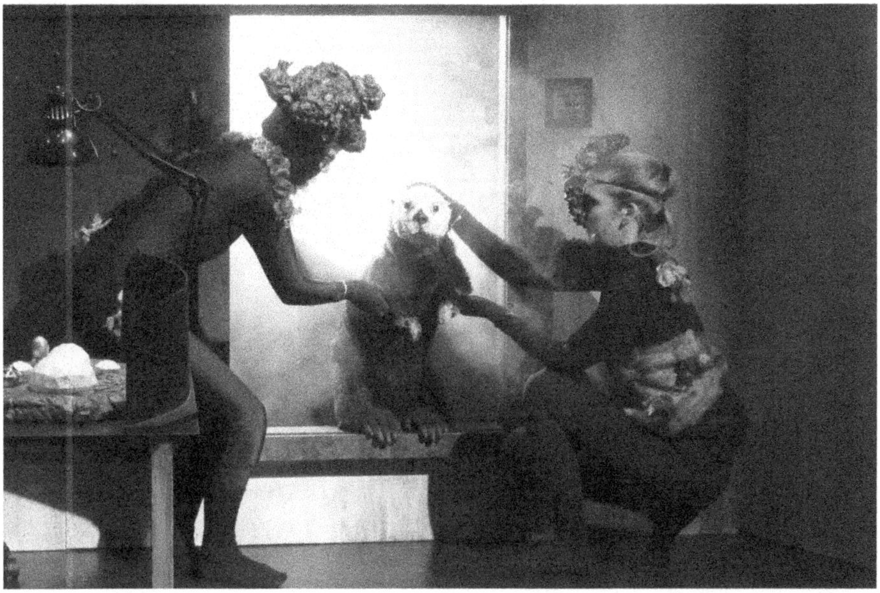

Figure 5.3 A Brief History of Human Extinction. Upintheair Theatre, 2018. Directed by Tamara McCarthy. Photo by Itai Erdal.

environmental disturbance. Donna Haraway's idea of "sympoiesis" underscores the importance of such collaboration as a key component of storytelling in the Anthropocene. For Haraway, sympoiesis means "making-with": "Nothing makes itself; nothing is really autopoietic or self-organizing" (*Staying with the Trouble* 58). Haraway uses the term "holobiont" to emphasise that an organism is never simply single or individual. The idea of the holobiont acknowledges "symbiotic assemblages, at whatever scale of space or time, which are more like knots of diverse intra-active relatings in dynamic complex systems, than like entities of a biology made up of preexisting bounded units" (60). By bringing the play into dialogue with symbiotic processes via the entanglement of Ommie and the fungal bodies, which can be seen as embodying Haraway's model of "tentacular" bodies, it is possible to say that humans cannot remain at the center of the play.[3]

The play's emphasis on collaborative stories offers alternative evolutionary models of community and kinship during the current global disequilibrium – models that are opposed to individualism and preexisting forms of relationship. As the playwright argues, "[m]oving away from really linear storytelling and into this more collaborative and imagistic approach does seem to be where we are going with telling these kinds of eco stories" ("Anthropocene Angst" n. pag.). The following conversation between Ever and Ommie/The Fungus is an example of multispecies collaboration, where we are intimately and dynamically entangled with strangers:

OMMIE/THE FUNGUS: I'm not strange. . . . You would think of me as Kingdom: Fungi, Genus: Cryptococcus –

EVER: Fungi. Cryptococcus. (*A beat.*) You're – you're D3. . . . I'd rather you hadn't killed my friend and used her corpse to trick me.

OMMIE/THE FUNGUS: The Otter You? She's not dead. . . . Why would I do that?

EVER: It's what D3 does.

OMMIE/THE FUNGUS: I'm not D3.

EVER: You killed off my species and you want to quibble about nomenclature?

OMMIE/THE FUNGUS: How would you like it if I went around calling you Australopithecus?

EVER: Australo –

OMMIE/THE FUNGUS: That's the relationship. Australopithecus evolved into you. Your "D3" is evolving into me.

EVER: So you're . . . D4? . . . How do I know you're not killing her?

OMMIE/THE FUNGUS: Well, for one thing, it would be so wasteful. If you kill a host, you can only gestate where it drops. Keep a host alive and you can shed spores wherever it travels. . . .

EVER: I don't really like the thought of something else controlling me.

OMMIE/THE FUNGUS: Not controlling. Communicating. Chemically. (*Beat.*) You're awfully stubborn. . . . I suppose. You are . . . part of me. The translation would be . . . imperfect . . . (Hall, "Human Extinction")

The play invites the audience to reconceive of their relations to nonhuman participants in the ecological community (see fig. 5.4). While Ommie/The Fungus makes an argument for its/their inclusion with the samples being launched in the Ark, she tries to convince Ever that "Evolution does its best work in collaboration" (Hall, "Human Extinction"). Ommie/The Fungus defines collaboration as what "you have here with all the other Yous – they feed you and clothe you and keep you company. . . . Viruses you housed gave you mitochondria. Placentas. Who knows what we could become in your mind, with all those lightning fast neurons. All these gorgeous words" (Hall, "Human Extinction"). The play uses the kind of performative and collaborative language considered necessary for raising awareness about how to survive in an ongoing ecological crisis. This possibility, of course, requires us to question the very idea of bounded individualism by asking "how humans need to reenvision their own identity so as to be able to live in societies populated by aliens who are at the same time intimately or even genetically related to humans – both self and other" (Heise, *Imagining Extinction* 228). In the play, this scenario of evolutionary – but messy and complicated – relationships between humans and nonhumans is imagined through the flexibility, materiality, and collaborative nature of the theatrical performance itself. In the production by Mind of a Snail in 2018, the stage's backdrop consisted of enchanting hand-drawn scenes

Figure 5.4 *A Brief History of Human Extinction*. Upintheair Theatre, 2018. Directed by Tamara McCarthy. Photo by Itai Erdal.

that pictorialise the real world poisoned by ominous fungal forms and a threatening multi-eyed creature. The overlapping shadows of the actors and puppets, shadow theatre backdrops, and skilfully multifaceted projections creatively materialise the entanglement of humans and nonhumans with the changing environment.

The evolution of a dynamic process of communication beyond the human, with all manner of other critters and forces, cultivates a democratic multispecies narrative via theatrical performance. Anthropogenic interventions in ecosystems reveal the illusion of the earth as a geologically stable entity and problematise material agency while nonhuman matters such as bacteria and fungi change into agential forces generating tentacular webs of both creative and destructive relations in an unpredictable way and beyond human imagination. Haraway argues that sympoietic assemblages " 'do not have self-defined spatial or temporal boundaries' " and they " 'are evolutionary and have the potential for surprising change' " (*Staying with the Trouble* 61). Such unpredictability is evident in Ommie/The Fungus's statement that they have potentially dramatic impacts, growing – as the spores of some fungal species might – by "spreading out in a glittering shower. Never knowing what will happen to all those millions upon millions of little possibilities" (Hall, "Human Extinction"). The possibilities arise, as Ever indicates, from the "speed" and "flexibility" that human interventions have, intentionally or unintentionally, given to nonhuman matter. The following conversation between Ever and Ommie/The Fungus further highlights the complexity emerging from the agency of nonhuman forms:

OMMIE/THE FUNGUS: You don't know what else we can do. I don't know what else we can do. We're something new.

EVER: And you're either a benign consciousness offering symbiosis, or a terrifying mind parasite in the process of turning everything it touches into zombies. . . .

OMMIE/THE FUNGUS: You're awfully preoccupied with control.

As Ommie speaks, the camera (now controlled by the fungal puppeteers) comes to point at Ommie, now speaking to us, much as Ever did at the outset of the play. With a gesture, the evolutionary tree reappears on the screen, but instead of travelling up, it scrolls down, into the mycelial world.

OMMIE/THE FUNGUS: Yes. The price to send all those little possibilities on their way. But you don't know anymore. You're so scared they'll have minds like yours. They'll break the world into pieces and make a pile so they can sit on top. If you let me, I can show them how it is all always connected. Your feet become the earth, the earth becomes the trees, the trees become the birds, the birds become the sky. The flux of all of us breathing and pulsing together. All those little Yous won't be alone inside themselves where they can forget that. They will be birds and fish and flowers and otters. They will carry this memory of you with them – not a handful of sounds and images, but all the glorious patterns of your knowing in this moment. A true evolution. (Hall, "Human Extinction")

This conversation illustrates that diverse nonhuman players – the whole earth, trees, birds, sky, fish, flowers, otters, fungi – might all be considered agential forces and selves that produce new meanings and narratives through collaboration with each other and with humans. In these complex processes, matter, in all its forms, becomes " 'material narrativity' " and "storied matter," expressing its own creative ongoingness and productivity (Iovino and Oppermann 1, 8). Seen in this light, as Serenella Iovino and Serpil Oppermann argue, "every living creature, from humans to fungi, tells evolutionary stories of coexistence, interdependence, adaptation and hybridization, extinctions and survivals. Whether perceived or interpreted by the human mind or not, these stories shape trajectories that have a formative, enactive power" (4) in the face of ongoing ecological crisis.

The play's narration and representation of such complex evolutionary movements offers a way of learning to "stay with the trouble of complex worlding," to use Haraway's expression, as an alternative model of living in a damaged world. As Haraway argues, "[s]taying with the trouble requires making oddkin; that is, we require each other in unexpected collaborations and combinations, in hot compost piles. We become-with each other or not at all" (*Staying with the Trouble* 4). The strange and complicated figure of Ommie/The Fungus and its ongoing tentacular troubling of the human characters is the result of such an oddkin process, which turns onto a new path by creating unexpected connections. As the following conversation exemplifies, the agency of all human and nonhuman critters is central to sympoietic collaborations and generative oddkin relationships in our current troubled times:

EVER: You have some troubling tendencies.
OMMIE/THE FUNGUS: Who doesn't. We can help each other with those. Maybe. If you let us become ourselves. (Hall, "Human Extinction")

In this agential material network, the collaborative survival of oddkin offers the only chance of staying with the trouble. To embrace Ommie/The Fungus as oddkin expands our understanding of the importance of shaping unexpected connections with the environment's apparently invisible and weird forces. Fungal oddkin make collective life on the earth possible by serving a double function in their interactions and mutual associations with other critters. While fungi decompose human flesh and can cause toxicity and infection, they also play an essential role in ecosystems by facilitating the cycling of nutrients and breaking down organic materials into simple molecules. In this way, thinking of fungus as oddkin challenges the binary of survival versus death because material decomposition does, indeed, enable the continuation of earthly existence and suggest a multispecies future that puts an end to human exceptionalism. For Haraway, the sympoietic process of living and dying together – or, in other words, "collective thinking and movement in complexity" (29) – constitutes the new era of Chthulucene. The Chthulucene is a useful term for remaining in the present epoch of ecological crisis. Haraway's Chthulucene derives from the Greek word *khthonios*, or of the earth (173), and is "made up of ongoing multispecies stories and practices of becoming-with in times that remain at stake" (55). The ideas of decentering the human subject and becoming-with all life on the earth are enacted in the play's collaborative narrative and communal performance, which demonstrates a chthonic model of life. This model features the material entanglements of earth processes and critters, however messy and complicated those are in the face of unexpected environmental changes.

In a chthonic reading, the play invites us to "walk together. Out into a new world" (Hall, "Human Extinction"). Ommie/The Fungus asks Adam to help rebuild the garden: "We will be the birds of the air, and the beasts of the field and together we will rebuild the garden. . . . The garden as it was in the beginning. With no master and no names. No beginning and no end. Life, forming and reforming endlessly" (Hall, "Human Extinction"). However, in line with the play's biblical allusion to the Garden of Eden, Adam sees this creature as repulsive and its suggestion as horrifying and deceptive, calling it the "serpent in the garden" that tries to control them. Adam, who is obsessed with bounded individualism, claims "I won't let it make me into something else" (Hall, "Human Extinction"). However, when he goes back into the burnt-out orchard enclosure, which is "*stretching out like a skeletal garden*," he is entangled with tendrils of the lethal D3 fungus: "*Adam tries to burn it with the flamethrower, but [the flamethrower has] run out of fuel. All around him, ash and spores drift through the air. He moves towards the skeletal apple trees. He collapses. [A] spore births itself from his chest*" (Hall, "Human Extinction"). This scene dramatises the intra-active entanglements of human and nonhuman actants in a chthonic world in which humans are not the only important participants. Rather than being autonomous, isolated individuals, humans exist inevitably in symbiotic assemblages, their agency is subject to ecological processes, and finally they end as compost.

Meanwhile, Ever begins the launch sequence – sending the new, entangled biosphere into the unknown – while she and the hybrid Ommie/Fungus "*move towards the world, until they are silhouettes and shadows*" (Hall, "Human Extinction"). In the 2018 production in collaboration with Mind of a Snail, shadow puppets and the shadows of live actors appear in colourful sets created through overhead projections (Thomas n. pag.). While the light fades on their silhouettes, we hear Ever's final message:

> There has always been a kind of war happening inside us. Between self-interest and cooperation, fear and bravery, cynicism and hope. I wanted to solve that for you – and I can't. You're alive and life is trouble. I can give you nothing except this chance: You will be like us and not like us, and that is the uncertainty that Hope is made of. Be better than us. Or worse. Or the same. Be the next moment in our story. Be. (Hall, "Human Extinction")

By bringing Ever's description of the uncertainty that emerges from unexpected exchanges between humans and the environment into dialogue with Haraway's sketch of the "disturbing," "mixed-up," and "troubling" times we live in (*Staying with the Trouble* 1), we can see that the play constitutes the kind of ecocritical theatre that can demonstrate how to persist in ongoing planetary crisis by making unexpected collaborations and combinations with, across, and in/between difference. In the play's troubled environment, survival and death are inextricably linked in the process of intra-actions in oddkin communities. Haraway argues that staying with the trouble involves the following challenges:

> Alone, in our separate kinds of expertise and experience, we know both too much and too little, and so we succumb to despair or to hope, and neither is a sensible attitude. Neither despair nor hope is tuned to the senses, to mindful matter, to material semiotics, to mortal earthlings in thick copresence.
>
> (4)

In an unprecedentedly troubled multispecies world, a radical and compelling way to stay with the trouble is to collaborate with other voices and search for emergent practices of "multispecies environmental justice of myriad kinds" (Haraway, *Staying with the Trouble* 8). The idea of multispecies kinship informs my reading of the play as an imaginative process of oddkin world-making that recognises the "thick copresence" of "mortal earthlings," and that defines death and life as closely connected to each other. This reading shows how the play imagines the possibility of generating various kinds of interrelations and a different future of co-becoming in a troubled environment.

Following Ever's message, the Ark "*launches out into the huge dark of Space, headed for Kepler. As it does, for the first time, the title card appears: A Brief History of Human Extinction*" (Hall, "Human Extinction"). Despite the fact that we are in the midst of a mass extinction crisis that could cause the end of human life (and the end of human exceptionalism) and the collapse of the natural ecosystems we rely upon to survive, the play highlights the importance of telling stories through these troubled times in order to be truly present. According to Haraway, "staying with the trouble requires learning to be truly present, not as a vanishing pivot between awful or edenic pasts and apocalyptic or salvific futures, but as mortal critters entwined in myriad unfinished configurations of places, times, matters, meanings" (*Staying with the Trouble* 1). Seen in this light, the play reminds us that surviving collaboratively in a toxic and contaminated environment is a skill that we need in the current ecological crisis. In contrast to Adam's pursuit of an "edenic past" and Ever's search for a new territory and a "salvific future," the play challenges the audience to think critically, provoking them to consider the possibility of staying with the trouble, which is "a commitment to living and dying with response-ability in unexpected company" (Haraway, *Staying with the Trouble* 38). This practice of learning to live and die with each other requires cultivating the capacity to be sensitive to, and create conditions for, ongoingness. Hall's play and its science fiction characteristics open paths for such practices of living and dying, acting, caring, and responding (response-ability) in a chthonic world of multispecies stories flourishing on a wounded planet.

2 The ecodramaturgical practice of sympoietic thinking in Miwa Matreyek's *This World Made Itself* and *Infinitely Yours*

Matreyek's works are intricate and embodied illustrations of the history of the Earth, the complex world of humanity, and the ecological demands of the Anthropocene. In her multimedia performances, including *This World Made Itself* (2013) and *Infinitely Yours* (2020), Matreyek does not follow a script written in advance, portray a specific character, outline a clear plot, or use straightforward language. Rather, she integrates multi-layered animated projections into live theatrical presentations, blurring the boundary between real and virtual. In particular, Matreyek combines the live shadow silhouette of her own kaleidoscopic movement with cinema, video projections, and recorded music. Her silhouette takes on the role of a creator, director, performer, and subject by moving behind the camera as well as in front of or among the animated images on the screen. As Holly Willis notes, "[t]he mix of projection and live performance produces a dizzying and captivating kind of vertigo as you try to trace what's real versus what's projected while marvelling at the magic of it all" (77). Matreyek thinks that it is important to use her shadow from visceral, physical, and abstracted perspectives, arguing that as a shadow, "my body can get larger than life, to show close-ups of my face, my hands, my feet, etc., as well as play with cinematic tropes and conventions. . . . By combining a real body with the video, I want the dreamlike or nightmarish visual moments to seem like real experiences happening to the physical body" ("BWW Interview" n. pag.). Further, her play with illusion and reality expresses complex visions of transgressing and dismantling boundaries between the body and the physical environment. She describes the merging of reality and illusion this way:

> I am certainly interested in the middle ground between the cinematic space and the staging, which I think my work is always in between. My body becomes transformed and becomes a bit more fantastical whereas the animation feels a bit more tangible and real because of the intersection with my body. I'm interested in breaking apart the cinematic space and making it a little bit more visceral, as well as have a feeling of consequence to a real body.
>
> ("Hunger" n. pag.)

Willis also describes the immediacy of Matreyek's physical and virtual body: "The body is at once entirely physical, but also subject to projection and reconfiguration via the media with which it interacts" (80). I argue that the flexibility and dynamism of Matreyek's works result in spatial-temporal distortions and create a collection of stories about the material world and its sympoietic connection to human bodies that are not only imaginary, fantastical, visual, and illusory, but also tangible, intimate, scientific, and physical.

As I show in the following sections, the disturbing merger of illusion and reality in Matreyek's performances evokes an emotionally and materially effective representation of complex interdependency in environmental disturbance by breaking the Anthropocene illusion that human agency is the driving force of the world. Her integration of art forms embodies a tension between the illusion of human subjectivity and the reality of sympoiesis that restricts and reconfigures human mastery over the earth's environments. These art works show that we are not only planet destroyers but inevitably and fundamentally affected by what is being changed, poisoned, and destroyed. On the one hand, the scope of anthropogenic impact on the environment invites us to extend our conception of the human as a geological agent. On the other hand, because of the unintended consequences of industrialisation and human actions, humanity's geological agency puts limits on human freedom. The ways in which Matreyek's multi-media performances represent the world exemplify how living in the Anthropocene means reconsidering cause-and-effect relationships because the complexity of climate change and mass extinction – together with the agency of all kinds of energy and matter – exceeds the intentionality of the humans who are causing it, trapping them in an interconnected ecological system.

An embodiment of the earth's deep time in This World Made Itself

This World Made Itself premiered at the Wexner Center for the Arts in Columbus, Ohio, in November 2013. This performance stages the history of our planet, evolution, and the complex modern world of humanity by invoking a simplified natural history and science reminiscent of flipping through a children's encyclopedia to create a dream-like, poignant narrative rich in surrealism, metaphor, and fantasy. This work was originally inspired by Matreyek's obsession with science and natural history museums with their dioramas of a prehistoric world and moments of evolution. In the first half of the performance, Matreyek's shadow silhouette takes a visual and musical journey through the universe's first spark, the formation of the world, and then compelling experiences of different stages of evolution recorded in geological strata: Pre-Cambrian life in the oceans, Ordovician marine life, and Carboniferous landscapes and forests. This journey involves incredibly colourful and surreal animated images of molten rock, the formation of the atmosphere, and the first oceans and landmasses, along with the emergence of proteins, which are built from chains of organic molecules such as amino acids that make up all living things and were the earliest biological processes on earth. In all these scenes, a human shadow is integrated with music and animated imagery to answer the question, "what does it feel like to be the earth that's just forming?" (Matreyek, "A Fantastical Interpretation of the Earth's Formation" n. pag.). The shadow body awakens as a molten rock floating in a lava flow. This scene is followed by Matreyek's shadow swimming in oceans and forming marine life with her mouth and hands, walking through grasses while a forest grows from her steps, and wandering across mountains (see fig. 5.5 and 5.6). These scenes show the shadow as a representation of the earth interceding in the process of evolution and prehistoric life. Rather than considering the shadow a specific person, Matreyek argues that the "[s]hadow can become a vessel for the audience to experience the world of what is happening in the performance" ("A Dreamy Shadow" n. pag.) as well as to imagine the human body as co-extensive with evolution and the earth's changes over time.[4]

The other source of inspiration for this work was Bill Bryson's book *A Short History of Nearly Everything* (2003). It is devoted to an enlightening overview of scientific thinking on everything from the Big Bang and the creation of the universe to the discovery of quantum mechanics, the history of geology and biology, the development of modern *Homo sapiens*, and our impact on the earth's climate and ecosystems. Matreyek's work is also inspired by the landscapes that she, as a touring performance artist, has seen from airplane windows. Comparing natural landscapes and engineered ones, Matreyek seeks a scientific, physical, visceral, and emotional way of reading and connecting with landscapes and their formation throughout the earth's deep-time history.

Figure 5.5 *This World Made Itself.* The Wexner Center for the Arts, Columbus (Ohio), 2013. Directed and performed by Miwa Matreyek. Photo courtesy of the artist.

Figure 5.6 *This World Made Itself.* The Wexner Center for the Arts, Columbus (Ohio), 2013. Directed and performed by Miwa Matreyek. Photo courtesy of the artist.

Painful pictures of anthropogenic fault lines, dead riverbeds, and open-pit mining and mineral processing plants that come from looking out of airplane windows are reflected in open sores and scratches on the human body in the performance (Matreyek, "Process and Inspiration"). The evolution of life on the earth in the first half of the performance stands against the history of the modern world and the urban-industrial expansion of human society spreading before the performer's silhouette in the second half. Matreyek's figure moves through projected images, dramatising the connection between environmental degradation and the wounded human body by, for example, superimposing a scene of mountaintop-removal mining on the shadow of her head and skull (see fig. 5.7). The image of the mountain and the shadow of her profile overlap, and as the mountain is broken apart and taken away piece by piece, her skull and her whole body are stripped bare and begin crumbling (Matreyek, "Process and Inspiration"). Matreyek sketches interdependent ecological relationships by imagining her body as the earth and putting her body into fractured landscapes. The performance prompts us to reflect on the illusion of human individuality and ask how humanity's devouring of landscapes affects human mind and flesh itself.

A different scene invites the audience to think about food imagery and food policies across historic periods. In this scene, the food on a dinner table

Figure 5.7 This World Made Itself. The Wexner Center for the Arts, Columbus (Ohio), 2013. Directed and performed by Miwa Matreyek. Photo courtesy of the artist.

Figure 5.8 This World Made Itself. The Wexner Center for the Arts, Columbus (Ohio), 2013. Directed and performed by Miwa Matreyek. Photo courtesy of the artist.

travels back in time to reveal how that food got there (see fig. 5.8). The food undergoes a complex set of material and transformative trajectories, recalling distant pasts and projecting futures. On the one hand, the growing demands for food, for example, have resulted in the conversion of natural ecosystems to crop production and pasture lands, followed by scenes of livestock on a factory farm and finally death at a slaughterhouse. By connecting the dinner table to these sequential images, the play shows that the food system is the driving force of deforestation, habitat loss, and biodiversity loss. On the other hand, the performance shows how all the life on earth is made out of recycled materials. The stuff that makes up our bodies used to be part of the food that we eat, whether it be animal or plant, which in turn was part of animals, plants, or soil (Matreyek, "Process and Inspiration"). Like Caryl Churchill's *The Skriker*, Matreyek's performance shows that food undergoes a long chain of transformations that can never be disentangled from the human body and mind. As Stacy Alaimo argues, "we are transformed by the food we consume . . . but for the most part the model of incorporation emphasizes the outline of the human: food disappears into the human body, which remains solidly bounded" (*Bodily Natures* 13). The participation of Matreyek's shadow in the sequence of animations about the food's transformations – from producer organisms to consumers and then to decomposers – merges the real and the surreal, extending the body outward into material interchanges with its chaosmos. In this way, connections between deep pasts and dark ecological futures caused by human actions "reside within everyday experience" in an intimate and compelling way (Ginn et al. 222). Here, the idea of planetary deep time is manifested through the agency of food, which is intertwined with daily life. Placing the history of the food and its trajectories in protracted temporal scales confronts human-centred conceptions of history, offering a way of thinking about our inevitable engagement with deep time. This interconnected engagement, in which "what is supposed to be outside the delineation of the human is always already inside" (Alaimo, *Bodily Natures* 25), is dramatised earlier in the performance by a scene of a supernova explosion that illustrates a scientific understanding of what human bodies are composed of. By zooming in on the cataclysmic explosion of a star at the end of its life and zooming back out, the performance shows us that stars are made of the chemical elements that compose our bodies and everything else in our universe. These elements include carbon, oxygen, nitrogen, and iron, which are required for life. Hence, as Alaimo asserts, "an understanding of the material interchanges between bodies (both human and nonhuman) and the environment often requires the mediation of scientific information" (16). By bringing art and science together, Matreyek's work shows how art may offer new ways of imagining the material world and biospheric interconnectedness.

However, this work suggests new ways of understanding the idea that anthropogenic activities – including population growth, overconsumption, burning fossil fuels, emitting greenhouse gases, overexploiting resources, reducing biodiversity, deforestation, ocean acidification, pollution, warfare, and oppression – have provoked the unexpected mutation of the elements which compose human and nonhuman bodies. For example, in one scene, we see the human shadow suddenly changes to a bird which is then decomposed and converted into a rocket. The rocket enters Earth's atmosphere, followed by an explosion and the destruction of the whole land. The second half of the performance includes a series of ecological mutations that shift our perception of the earth from the mythic story of "Gaia" as a single organism evolving over billions of years to Latour's contemporary concept of it as "an entity composed of multiple reciprocally linked but ungoverned self-advancing processes" ("Facing Gaia" 8). Bruno Latour evokes the historic idea of Gaia, initially developed by James Lovelock, expanding it from a "figure of unification" to a "name proposed for all the intermingled and unpredictable consequences of the agents, each of which is pursuing its own interest" (*Facing Gaia* 142). I argue that this way of re-reading the model of Gaia helps identify how the performance dramatises human agency by situating it in an endless network of relationships to prompt engagement with what has been done in the past and with the future's possibilities (see fig. 5.9). The integration of the shadow of the human body, performance, and animation offers the audience an opportunity to face Gaia and explore invisible, complex ecological and biological changes that happen at different temporal and spatial scales. Matreyek's artwork allows us to see the world emotionally and materially in a way we never have before. As I show in

Figure 5.9 This World Made Itself. The Wexner Center for the Arts, Columbus (Ohio), 2013. Directed and performed by Miwa Matreyek. Photo courtesy of the artist.

the following section, the story of Gaia and deep-time history in this work is completed in Matreyek's subsequent play, *Infinitely Yours*.

Wandering in the Chthulucene in Infinitely Yours

Infinitely Yours premiered at the Roy and Edna Disney/CalArts Theatre (RED-CAT) in Los Angeles in January 2020. Matreyek's silhouette shapeshifts to a live music soundtrack performed by Morgan Sorne, known in his solo music project as SORNE. Her shadow traverses manifold scales and materialises the dramatically changing world from the perspective of different consciousnesses and collective experience. Inspired by the Anthropocene and Matreyek's own climate grief, the dreamlike world of this layered-projection performance portrays a threatening earth overflowing with climate disasters, unprecedented wildfires, storms, massive floods, sea level rise, droughts, famines, pollution, and contamination. The main sources of inspiration for this interdisciplinary stage performance were news items, articles, and headlines about climate change and environmental transformation that Matreyek has been reading over the last few years (Matreyek, "BWW Interview"). She has catalogued familiar media reports and turned them into visual images and performative narratives of embodied experiences, yielding a cacophony of dreamlike, emotional, and physical representations of the interrelationship between action and effect, subject and object, and the human and material world. The result is a diversity of images that evoke a multi-dimensional environment as seen from different perspectives, such as that of the sky, solar system, land, underground, underwater, and inside and outside the human body. The shadow of the performer interacts with these images and travels across them like a shadow puppet, shapeshifting from an observer to a creator, destroyer, and one who is being destroyed as she faces her own climate anxiety.

A major part of the imagery in the performance also originates from Matreyek's daily life, since she uses the everyday plastic trash she generates to create material scenes in which these plastics float in the polluted ocean (Matreyek, "BWW Interview"). This work forces Matreyek and the audience to confront their own personal choices and responsibility as consumers who pollute the environment and change lifeforms in it with all of the plastic waste and other indestructible objects that they purchase and discard in their own daily lives. The performance represents human activities that have dramatically transformed the course of world history through images, such as that of a huge body rising along with a city, traversing urban sprawl and industrial areas, and destroying forest and clearing land. These images are accompanied by scenes of animal carcasses hanging in a slaughterhouse, plastic chain conveyor belts of bottled water at factory, mass production, drilling for and extracting fossil fuels and natural gases, fish caught in a trawling net, and masses of dead sea creatures. As the performance moves on, such images turn into ones of landfills of plastic, oil spills, and droughts.

As the title *Infinitely Yours* implies, this performance takes on an evolutionary narrative about the entanglement of anthropogenic activities, nonhuman species, climatic and natural forces, and the human body. The work begins with a local manifestation of climate change in California that is represented through scenes of forest fires against an orange sky. Matreyek's silhouette, in live performance, battles through intense, destructive blazes burning trees and grasses. In an email conversation with Matreyek, she asserts that this performance feels "California inspired in many ways, with many concerns and landscapes being incorporated from the world around me." For example, the ever-present worry about wildfires, the Los Angeles River, and the oilfields of Central California have led different scenes of the performance (Matreyek, "New Form"). However, she indicates that this image in the opening scene obviously embodies the same kind of devastation in other parts of the world, including the uncontrollable rainforest fires in the Brazilian Amazon and the catastrophic wildfires across Australia ("Prix Ars Electronica" n. pag.). In the following scene, a crack opens in the ground and buildings emerge from it, while the shadow's giant body also comes out of the ground. Accelerating anthropogenic ecological problems have buried oceans and lands under increasing layers of plastic waste and toxic chemicals, altering not only the earth's surface processes but also the human body. To illustrate the vast effects of producing and consuming plastics, Matreyek steps behind the screen to work alongside her animated worlds, dramatising a series of interactions between humans and the environment: a giant body digs into the ground as oil pump jacks do, extracting crude oil (the earth's blood) from oil wells (the earth's heart), pumping it out of the ground, and gobbling it up, while the oil suddenly becomes blood flowing out of her body. Her shadow also tosses plastic trash in a landfill and drowns in a plastic-filled ocean (see fig. 5.10, 5.11, 5.12). In this industrial environment, contaminants penetrate and build up in the body in many ways, from eating a fish that is exposed to plastic toxins to drinking contaminated water and breathing polluted air. Eating plastic-contaminated fish threatens to poison the shadow's body. The oil spilling from a pipeline contaminates a river that Matreyek's shadow drinks from. The air also becomes polluted with toxic fumes and chemicals emitted by a gas station, which are absorbed by her shadow. The silhouette's toxic body, with damaged lungs, is a manifestation of the devastating traces that the threatening forces of modernisation and environmental exploitation leave in us (see fig. 5.13). More broadly, as the performance shows, her damaged lungs represent the world's devastated forests and oceans, which are referred to as the lungs of the earth. While the shadow's lungs draw in enormous amounts of the carbon dioxide that circulates in the atmosphere, the performance reveals how her lungs reflect deforestation and degradation. The shadow's body then grapples with a violent thunderstorm, falling trees, and a river's rising waters.

Figure 5.10 *Infinitely Yours*. The Roy and Edna Disney/CalArts Theatre (REDCAT), Los Angeles, 2020. Directed and performed by Miwa Matreyek. Photo by Keida Mascaro. Photo courtesy of the artist.

Figure 5.11 *Infinitely Yours*. The Roy and Edna Disney/CalArts Theatre (REDCAT), Los Angeles, 2020. Directed and performed by Miwa Matreyek. Photo by Keida Mascaro. Photo courtesy of the artist.

Figure 5.12 *Infinitely Yours*. The Roy and Edna Disney/CalArts Theatre (REDCAT), Los Angeles, 2020. Directed and performed by Miwa Matreyek. Photo by Keida Mascaro. Photo courtesy of the artist.

Figure 5.13 *Infinitely Yours*. The Roy and Edna Disney/CalArts Theatre (REDCAT), Los Angeles, 2020. Directed and performed by Miwa Matreyek. Photo courtesy of the artist.

Rising sea levels and rushing currents overflow and submerge buildings and houses. Meanwhile, floating plastic waste inundates lands and oceans, flooding the human body inside them.

The shadow passes between the human body and the body of the earth to illustrate that humans are inescapably part of a larger ecosystem but vulnerable to anthropogenic ecological change. As the body of the earth, the shadow shows how cosmic balance undergoes change that outpaces our ability to control it. As a human body, the shadow tells us that our bodies can never be disentangled from biospheric changes, as the self is deeply affected and constituted by the food that we eat, the oxygen that enters our lungs and diffuses into the blood, and other material agencies. Hence, "when we contaminate the planet we contaminate ourselves" (Matreyek, "Process and Inspiration"). The body's porous boundaries immerse it in a drastically mobile environment, which, "as we now apprehend it, runs right through us in endless waves, and if we were to watch ourselves via some ideal microscopic time-lapse video, we would see water, air, food, microbes, toxins entering our bodies as we shed, excrete, and exhale our processed materials back out" (Fromm 95). This quote from Harold Fromm's "The 'Environment' is Us" gives a remarkable image of Alaimo's concept of "trans-corporeality," implying the transit of material substances, such as plastic particles and toxins, across human bodies and the environment over time and space. For example, given that the average American produces 4.4 pounds of trash a day, Matreyek asks: "What does it mean when you multiply that by a week, a month, a year or a lifetime? Or, a whole household? What would it feel like for all that trash to come pouring out of your body?" ("Process and Inspiration"). I argue that, in the performance, trans-corporeality defines the human as a material entity inextricably connected with and constituted by nonhuman natures, with ethical and political implications for how we – as individuals and collectivities – act in relation to the ever-changing world. Matreyek's innovative strategy of enacting a collaboration between her shadow and projected animations results in a visualisation of the vast scale of "slow violence" as a personal, embodied experience, opening up a trans-corporeal understanding of the relationship between our bodies and "landscapes of strange agencies" and invisible threats (Alaimo, *Exposed* 78) that are caused by anthropogenic damage to the planet.

Throughout the performance, the collaboration of the shadow's movements with the solo music project SORNE – known for its five-octave vocal range – creates more effective metaphoric and physical images of the familiar headlines we see every day, acting as a sonic illustration of the alterations we have made to our planet and, consequently, to our bodies. This visual, rhythmic, and tactile space – picking up where *This World Made Itself* left off – asks us to take into account the vast scale of deep time in narrating human stories in the Anthropocene. Matreyek's attention to the immediacy of the current ecological crisis fosters connections with deep time by making visible and tangible manifold entangled bodies and processes and their interrelationships across time and space. To reveal the entangled alliances between human bodies and a multitude of nonhuman others and to show how experiences of climate change directly affect those bodies, Matreyek integrates her own body in shadow into her animations. As she argues, "I was interested in sudden flips and who the shadow figure might represent. Some of the moments I have an individual person experiencing the world, but sometimes my shadow represents all of humanity and yet again sometimes the earth itself. My shadow flips between being a destroyer and the destroyed" ("Process and Inspiration"). Matreyek performs the world so as to demonstrate how each body and component is in sympoietic relationship with a messy and complicated world of materiality. Thus, her work reflects our historical moment of biological, ecological, and cultural confusion, resonating with Haraway's idea of the Chthulucene as a model for thinking about the tentacular, entangled, and trans-corporeal material world. As a complement to Latour's idea of Gaia, a chthonic perspective indicates that "human beings are not the only important actors in the Chthulucene, with all other beings able simply to react. The order is reknitted: human beings are with and of the earth, and the biotic and abiotic powers of this earth are the main story" (*Staying with the Trouble* 55). I argue that the performance tells the story of the Chthulucene by dramatising a collective ecological system that has no predictable, determinate spatial and temporal boundaries, but posits the entanglement and embeddedness of humanity in the dynamism of the living earth.

As the performance proceeds to its end, the shadow – as a representation of the human body and the body of the planet – enters the earth's core and remains a part of interweaving stories of creation and destruction, evolution and extinction, offering a glimmer of hope and a possibility of renewal amid ruins and precarity. Seeds crack open to sprout, plant roots extend their tendrils underground, and vegetables and flowers grow from seeds, shaping the planet and creating a breathable atmosphere for all creatures to thrive in. In the womb of the earth, a newborn baby is cradled in a bed filled with plants, mushrooms, and butterflies. However, the planetary cycle of creation turns to traumatic destruction and ecological meltdown, accompanied by imagery of forest degradation; the rise of buildings and towers on the planet's rim; oil being pumped up from deep underground or, as the performance shows, from the shadow's heart and flesh; tonnes of plastic pollution and tires falling; and fighter aircraft flying (see fig. 5.14). These devastating scenes lead to a massive explosion that creates a mushroom cloud, painting a dystopian picture of the end of the world. The continual pattern of life and death concludes with the scene of fire that we see at the beginning of the performance, reminding us of the stage of "dark ecology" in the final scene of Churchill's *Far Away*, in which humans, nonhuman

Figure 5.14 Infinitely Yours. The Roy and Edna Disney/CalArts Theatre (REDCAT), Los Angeles, 2020. Directed and performed by Miwa Matreyek. Photo courtesy of the artist.

species, natural forces, and all kinds of matter are affected by each other in a chaotic ecology.

On the one hand, out of the ruins of human activities comes a newborn baby, a prerequisite for the continuation of human life on earth, who might be seen as representing a sense of possibility or a practice of what Anna Tsing calls "arts of noticing," which are about the necessity of understanding survival in the increasingly uncertain presents we inhabit (18–9). In her book *The Mushroom at the End of the World* (2015), Tsing focuses on collaborative survival and the "possibilities of coexistence with environmental disturbance" in order to stay with the precariousness of life under capitalism (4). On the other hand, depicting a newborn baby in the midst of all of this environmental tumult and transformation demonstrates the transference of pollution and toxicity from one generation to the next. Therefore, I argue that the story of the baby exemplifies the disruption of spatial and temporal boundaries and the extension of the concept of agency from humans to trans-corporeal matter and sympoietic nonhumans. As in Churchill's *The Skriker*, in which future generations of dead and malformed children are unknowingly affected by environmental toxicity, Matreyek shows that earthly processes occurring in a toxic world outlast generations and entangle diverse temporalities and spatialities. According to Tsing, existence is not self-defined and bounded but composite and entangled: "The evolution of our 'selves' is already polluted by histories of encounter; we are mixed up with others before we even begin any new collaboration" (29). This idea implies that "the modern human conceit is not the only plan for making worlds: we are surrounded by many world-making projects, human and not human" (Tsing 21), and our bodies are open to the unexpected influence of ecological changes. By interweaving her own body with those of so many others, I argue that Matreyek introduces a new mode of ecocritical performance focused on an emerging state of being composed of multiple narratives of radically invisible but material agents and climatic forces triggered by massive human activities.

Hall and Matreyek's works in this chapter dramatise that our planet is approaching a series of climate tipping points and mass extinction, and that the anthropogenic transformation of the earth system endangers all life forms including us. These works also remind us that humans can still avert the worst effects of climate change with drastic and immediate cuts to global emissions. This change, of course, requires a new perspective on our place in the world and a broadening of our frameworks of intimacy to include myriad nonhuman others. As Haraway argues,

> The decisions and transformations so urgent in our times for learning again, or for the first time, how to become less deadly, more response-able, more attuned, more capable of surprise, more able to practice the arts of living and dying well in multispecies symbiosis, sympoiesis . . . on a damaged planet, must be made without guarantees or the expectation of harmony with those who are not oneself – and not safely other, either.
>
> (*Staying with the Trouble* 98)

The performances discussed here invite us to be more sensitive and response-able to the entangled beings of the Chthulucene. Hall shows the necessity of communicating the ideas of multispecies sympoiesis, making together, and staying with the trouble of tangled worlding by representing the unexpected collision of the last human beings on the earth, the last living sea otter, and the new species evolving out of an environment drastically altered by anthropogenic interventions. While Hall illustrates the urgency of this intricate situation by integrating science fiction into performance art, the participation of Matreyek's silhouette in the webbed processes of planetary dynamics makes us aware of our agency as chthonic subjectivities. This complex subjectivity refers to inherently interconnected, endangered, and sensitized beings, rather than autonomous and self-defined individuals, characterised by material intra-actions across boundaries of human and nonhuman bodies and spatiotemporal scales. In this new materialist perspective, these evolutionary performances of co-existence call for thinking ecologically, provoking a new awareness of our bodily materiality, vulnerability, and capacity for radical change, flourishing, and recuperation, as well as for taking action in line with these new ways of surviving collaboratively in troubling times.

Notes

1 According to Ursula Heise, "Contemporary speculative fiction . . . defines itself in the tension between two different approaches to the present. From one perspective, the present appears as the past of imagined futures, and from the other perspective as an already if incompletely materialized future that makes palpable the obsolescence of the present. The Anthropocene . . . can usefully be understood as the second kind of speculative fiction, in that it focuses on the reality of a terraformed planet that the genre has long held out as a vision for the future of other planets, but which has already arrived in the present on our own planet" (*Imagining Extinction* 219–20).

2 By integrating and interweaving the puppetry into the action, Chloé Ziner, the founder of Mind of a Snail Puppet Company, says that we "arrive at this interwoven multimedia spectacle play. . . . It's kind of like a graphic novel come to life with the imagery and the story" (Derdeyn n. pag.).

3 Donna Haraway categorises fungi tangles, along with creatures such as spiders, jellyfish, cnidarians, etc., as tentacular figures (*Staying with the Trouble* 32).

4 Inspired by the ecological philosopher Joanna Macy and her book *World as Lover, World as Self* (1991), Miwa Matreyek links the idea of co-existence of humans and nonhumans in her own work. As Matreyek mentions, Macy "talks about the idea of the greening of the self. If someone is cutting off their leg you'd say, 'Don't do that, you're hurting yourself.' Similarly, if someone is cutting down the rainforest, we need to see that as an extension of self. She said, 'The trees in the Amazon rain basin. They are our external lungs. We are beginning to realize that the world is our body.' Macy's ideas explore expanding your consciousness to more than your body into something that encompasses these larger ecologies that support you. Macy described the greening of the self as the process of replacing one's 'skin-encapsulated ego' with 'wider constructs of identity and self-interest – by what you might call the ecological self, co-extensive with other beings and the life of our planet." ("A Dreamy Shadow" n. pag.)

References

Alaimo, Stacy. *Bodily Natures: Science, Environment, and the Material Self.* Bloomington: Indiana UP, 2010. Print.

———. *Exposed: Environmental Politics and Pleasures in Posthuman Times.* Minneapolis and London: University of Minnesota P, 2016. Print.

Bryson, Bill. *A Short History of Nearly Everything.* New York: Broadway Books, 2003. Print.

Colebrook, Claire. *Death of the PostHuman: Essays on Extinction, Vol. 1.* Ann Arbor: MPublishing, Open Humanities P, 2014. Web. 11 Aug. 2015. <www.openhumanitiespress.org/books/titles/death-of-the-posthuman/>.

———. "We Have Always Been Post-Anthropocene: The Anthropocene Counterfactual." *Anthropocene Feminism.* Ed. Richard Grusin. Minneapolis: Minnesota UP, 2017.1–20. Print.

Derdeyn, Stuart. "Anthropocene Angst Meets Puppets in A Brief History of Human Extinction." *Vancouver Sun*, Postmedia. 08 Oct. 2018. Web. 05 Jun 2020. <https://vancouversun.com/entertainment/theatre/anthropocene-angst-meets-puppets-in-a-brief-history-of-human-extinction>.

Fromm, Harold. *The Nature of Being Human: From Environmentalism to Consciousness.* Baltimore, MD: Johns Hopkins UP, 2009. Print.

Gartlan-Close, Katie. "Theatre Review: A Brief History of Human Extinction Is a Unique Collaboration." *Vancouver Presents.* Vancouver Presents, 14 Oct. 2018. Web. 10 May 2020. <www.vancouverpresents.com/theatre/theatre-review-a-brief-history-of-human-extinction-is-a-unique-collaboration/>.

Ginn, Franklin, et al. "Introduction: Unexpected Encounters with Deep Time." *Environmental Humanities* 10.1 (2018): 213–25. Web. 15 June 2021. <https://doi.org/10.1215/22011919-4385534>.

Hall, Jordan. "A Brief History of Human Extinction." 2018. Theatrical script.

———. "Storytelling Strategies in Eco-Theatre." *Artists and Climate Change.* Artists and Climate Change, 6 Sep. 2018. Web. 12 Feb. 2020. <https://artistsandclimatechange.com/category/jordan-hall/>.

———. "Chapter with Notes." Received by Mohebat Ahmadi. 31 Oct. 2021. Email Interview.

Haraway, Donna J. "A Cyborg Manifesto: Science, Technology, and Socialist-Feminism in the Late 20th Century." *The International Handbook of Virtual Learning Environments.* Ed. Joel Weiss, et al. Dordrecht, NL: Springer, 2006.117–58. Print.

———. "Anthropocene, Capitalocene, Plantationocene, Chthulucene: Making Kin." *Environmental Humanities* 6.1 (2015): 159–65. Web. 20 May 2020. <https://doi.org/10.1215/22011919-3615934>.

———. *Staying with the Trouble. Making Kin in the Chthulucene.* Durham, NC: Duke UP, 2016. Print.

Heise, Ursula K. *Imagining Extinction: The Cultural Meanings of Endangered Species.* Chicago: University of Chicago P, 2016.

———. "Science Fiction and the Time Scales of the Anthropocene." *English Literary History (ELH)* 86 (2019): 275–304. Print.

"Infinitely Yours." By Miwa Matreyek, directed by Miwa Matreyek, performance by Miwa Matreyek. Sundance Film Festival, 27 Jan. 2020, Los Angeles, CA.

Iovino, Serenella and Serpil Oppermann. "Introduction: Stories Come to Matter." *Material Ecocriticism.* Eds. Serenella Iovino and Serpil Oppermann. Bloomington: Indiana UP, 2014. Print.

Jones, Owain, et al. "Everyday Ecocide, Toxic Dwelling, and the Inability to Mourn: A Response to Geographies of Extinction." *Environmental Humanities* 12.1 (May 2020): 388–405. Web. 10 May 2021. <https://doi.org/10.1215/22011919-8142418>.

Latour, Bruno. *Facing Gaia: Eight Lectures on the New Climatic Regime*. Translated by Catherine Porter, Cambridge: Polity P, 2017. Print.

———. "Facing Gaia: Six Lectures on the Political Theology of Nature." Gifford Lectures on Natural Religion. Edinburgh, 18–28 Feb. 2013. Web. 16 Sept. 2015. <https://macaulay.cuny.edu/eportfolios/wakefield15/files/2015/01/LATOUR- GIFFORD-SIX-LECTURES_1.pdf>.

Macy, Joanna. *World as Lover, World as Self*. Berkeley, CA: Parallax P, 1991. Print.

Matreyek, Miwa. "A Dreamy Shadow Show in Downtown Durango: Miwa Matreyek Performs at DAC." Interview by Patty Templeton. *Durango's Ghost Car*. Ballantine Communications, 26 Oct. 2017. Web. 25 May 2021. <https://dgomag.com/contents/2411>.

———. "A Fantastical Interpretation of the Earth's Formation." Interview by Kate Torgovnick May. *Ted Blog*, Ted Conferences, 19 Jan. 2014. Web. 14 May 2021. <http://ideas.ted.com/miwa-matreyek-newest-surreal-performance/>.

———. "BWW Interview: Artist Miwa Matreyek Talks Infinitely Yours." Interview by Don Grigware. *Broadway World*. Wisdom Digital Media, 14 Jan. 2020. Web. 10 June 2021. <www.broadwayworld.com/los-angeles/article/BWW-Interview-Artist-Miwa-Matreyek-Talks-INFINITELY-YOURS-20200114>.

———. "Infinitely Yours – Process and Inspiration." *MIWA Matreyek*, uploaded by Miwa Matreyek in partnership with FirstWorks for the Virtual Learning Series, 2020. <www.semihemisphere.com/#/infinitelyyours/>. <http://first-works.org/education/firstworks-virtual-learning-series/>.

———. "Interview: Miwa Matreyek." Interview by Aine Carlin. *Hunger TV*. Hunger, 24 Mar. 2014. Web. 14 May 2021. <www.hungertv.com/feature/interview-miwa-matreyek/>.

———. "Miwa Matreyek's Virtual Vertigo." Interview by Holly Willis. *KCET Art Bound*. Public Media Group of Southern California, 02 Feb. 2014. Web. 14 May 2014. <www.kcet.org/arts/artbound/counties/los-angeles/miwa-matreyek-media-artist-redcat.html>.

———. "Prix Ars Electronica 2020 / Computer Animation: Miwa Matreyek." *Ars Electronica*. Ars Electronica, 25 June 2020. Web. 10 June 2021. <https://ars.electronica.art/homedelivery/en/prix-ars-electronica-computer-animation/>.

———. "From Submission - New Form - Permission Request." Received by Mohebat Ahmadi. 31 Oct. 2021. Email Interview.

Oliver, Kathleen. "A Brief History of Human Extinction Explores Planetary Doom in Two and Three Dimensions." *Georgia Straight*. Vancouver Free P, 15 Oct. 2018. Web. 10 May 2020. <www.straight.com/arts/1150901/brief-history-human-extinction-explores-planetary-doom-two-and-three-dimensions>.

"This World Made Itself." By Miwa Matreyek, directed by Miwa Matreyek, performance by Miwa Matreyek, Wexner Center for the Arts, 31 Oct. 2013, Columbus, OH.

Thomas, Colin. "A Brief History of Human Extinction: Barely a Whimper." Web blog post. *Mapping the Intuitive*. Colin Thomas, 12 Oct. 2018. Web. 25 May 2020. <https://colinthomas.ca/2018/10/12/brief-history-human-extinction-barely-whimper>.

Tsing, Anna Lowenhaupt. *The Mushroom at the End of the World: On the Possibility of Life in Capitalist Ruins*. Princeton, NJ: Princeton UP, 2017. Print.

Willis, Holly. *Fast Forward: The Future(s) of the Cinematic Arts*. London and New York: Wallflower P, 2016. Print.

6 Conclusion

Anthropocenic theatre: a stage for living in the present and imagining the future

The plays and performance practices explored in this book have yielded new insights into how a theatrical work can grapple with the challenges of the Anthropocene. The creative and communal potentialities of these artworks from diverse Western contexts lie in their offer of pivotal sites of communication about the Anthropocene as a scientific as well as an ethical, political, and cultural concern. This book has foregrounded the ways in which these works by Caryl Churchill, Stephen Sewell, Andrew Bovell, E.M. Lewis, Chantal Bilodeau, Jordan Hall, and Miwa Matreyek can be understood as the kind of "writing in the Anthropocene" that Deborah Bird Rose has called for ("Introduction: Writing in the Anthropocene" 87). Taking inspiration from Val Plumwood's call for "thinking about writing for the Anthropocene," Rose insists on the necessity of writing that holds potential for "shaking up our culture, and awakening us to new and more enlivened understandings of the world, [and] our place in it" (87). In effect, this book has shown what an ecocritical reading of Anthropocene writing has to offer, by analysing this often overlooked aspect of theatre studies. Such a reading enables us to explore the role of theatrical narrative and performance in our current time of crisis, in which humans find themselves always already entangled with nonhumans.

The playwrights and eco-artist studied in this book were selected because they share profound insights into the reorientation of theatre from anthropocentric to ecocritical drama. My discussion of theatre through this lens not only advances my argument about the rise of ecodramaturgical theatre, but also identifies a shift in theatrical narrative and form as it engages with the themes of ecology, climate change, pollution, extinction, and catastrophe. Although some theatrical works fall into this category more obviously than others, this study has focused on the contribution of all the plays and performances to reimagining the world from different points of view, identifying representational shifts towards interactions between humans and nonhumans, and situating, personalising, and embodying the presence of hyperobjects, the effects of extinction, and anxiety about sustainability for a broader understanding of the Anthropocene. These inquiries stem from the works' capacity to collapse what Una Chaudhuri calls the fifth wall of anthropocentrism. Breaking the fifth wall enables encounters between bodies to move beyond the traditional certainties about Nature and

DOI: 10.4324/9781003048749-6

Culture that undergird the humanist bias of theatre. Such a reconfiguration of theatre in the face of the challenges of the Anthropocene make a timely intervention into what Alan Read calls "the human laboratory" of theatre. Read's idea refers to the site of "the last human venue" or "theater in the age of urban modernity," in which "distinctions between humans and other animals are played out" (1, 3, 272). According to Read, the sense of the presence and newness of the theatre as the last human venue "was braided with a vivid sense of its *ending*," as "its time did not correspond with 'its times'" (272). Read takes into account that this time of climate change and environmental risk inaugurates the end of the world as humans knew it, disturbing anthropocentric ways of seeing and framing a dramatic form. Building on this idea, I argue that ecocritical theatrical narratives offer a means of unfolding stories of all kinds of material forms and bodies. For theatrical narratives to achieve this task, it is imperative to think about stories beyond human-centred drama; that is, plays must acknowledge that global ecological change has posed a problem for human subjectivity, and must shed light on the dynamic performativity and creative expression of nonhuman agencies and forces. By questioning the anthropocentric basis of theatre, I am seeking new ways to effectively communicate real-world debates about the urgency of environmental problems and their impact on humans and nonhumans, though it is difficult to capture the complexity of ecological risk and its temporal and spatial dissemination in theatrical space.

This book has brought together six Anglophone playwrights and one eco-artist; their works were written in different ecological contexts with overtly different dramatic languages, forms, modes of representation, and individual responses to environmental change. However, while the plays and performances are principally oriented to differing landscapes, they are, in fact, not simply narratives about their own ecological contexts. As Serenella Iovino asserts, "we are in a place, but the flows of energy, foods, money, pollutants, and other relevant matter and ideas mean that our being is the outcome of much more widespread and physically diffused systems" (*Ecology, Resistance, Liberation* 2). As I have shown, each play and theatrical performance stages this global, non-anthropocentric perspective by linking the ecology of a particular landscape with the planetary biosphere and by depicting an entangled network in which a wide range of meanings, processes, relations, actions, stories, spaces, forces, and matter make human and nonhuman life.

There are other shared themes and preoccupations that unite these theatrical works. Taken together, they demonstrate the ways in which material, emotional, psychological, cultural, and artistic reconnection with the natural environment posits a dark ecological picture. In other words, these plays contribute to the rise of an ecophobic kind of theatre and performance, to use Simon Estok's term, in which fear of the material world's loss of predictability becomes fear of its agential forces and dynamics far beyond human agency, control, and imagination. This ecophobic situation – which is derived from the adverse impact of human-induced environmental displacement and reflects the disordered realities of the Anthropocene – arises from the unexpected agentic capacity of nonhuman fairy-like figures and toxic food in *The Skriker*; of all life-forms and a range of material entities including nonhuman animals, plants, weather, water, light, and sound in *Far Away*; of weather, rain, sound, energy, and electricity in *When the Rain Stops Falling* and *It Just Stopped*; of Bolivian ecocide and Cambodian genocide in *Song of Extinction*; of polar bears, climate, weather, sea, air, and carbon emissions in *Sila*; of otter and fungi in *A Brief History of Human Extinction*; of air, water, climate, food, oil, plastic waste, nonhuman species, natural forces, and the whole earth in *This World Made Itself* and *Infinitely Yours*. Most significantly, all these works attempt to give expression to the materiality of nonhuman bodies, animals, things, and phenomena, not as scenic backdrops or metaphors for human situations, but as narrative agencies that are entangled with human characters and are capable of communicating the complexities of ecological change. Further, the theatrical examples considered in each chapter rely on analogies between individual deaths and specific conflicts, on the one hand, and the threat of climate risk and extinction, on the other, to make global environmental change concrete, palpable, and moving.

By situating these works within global and material approaches to ecocriticism in the context of the Anthropocene, I have asked what role theatre and performance can play in reshaping and enriching our understanding of the concepts of subjectivity, humanity, scale, agency, and an entangled ethics of multispecies flourishing. By expanding my definition of "nature" to include the global and agential aspects of environmental degradation, I have focused on the questions of scale and agency, which are key to this study and which orient my research. I have discussed the materiality of scale, ranging from the planetary in Ursula Heise's concepts of eco-cosmopolitanism and deterritorialization and Timothy Clark's theory of the derangements of temporal and spatial scales, to the agential configuration of scale framing in Timothy Morton's idea of the hyperobject, Bruno Latour's concept of compositionism and shift from the global to loops, and Karen Barad's reconfiguration of space-time-matter. These theoretical interventions have enabled me to show how these plays and performances rethink and reframe critical classifications of scale that consider only images and narratives on the human scale, and instead dramatise a sense of dissemination, transition, and continuity. Drawing upon these theoretical concepts prompts us to consider these theatrical works in terms of Dipesh Chakrabarty's assertion that, in response to the Anthropocene, we need to "think human agency over multiple and incommensurable scales at once" ("Challenge of Climate" 1). This perspective compels us to reflect on the fact that the boundaries between the human and the geological are in crisis as anthropogenic interventions in the earth's natural systems and climate begin to materialise as the transformative power of nonhuman agents on a planetary scale. The works' representation of the mutual implications of personal, local, and global scales conveys a sense of the mutability and complexity of scale, which necessitates thinking beyond a human-centred narrative of the world and, instead, acknowledging the intersection of human and nonhuman scales in narrating stories of life in the current environmental crisis.

In order to illuminate how the plays and performances embody the entanglement of the agential forces of chemical toxins, fungi, carbon emissions, and extreme weather with nonhuman animals and human lifestyles and subjectivities, I have also invoked Stacy Alaimo's concept of trans-corporeality, Jane Bennett's vital materialism, Donna Haraway's concepts of sympoiesis and the Chthulucene, Latour's actor-network theory, and Barad's concepts of intra-action and posthumanist performativity. These theoretical concepts from material ecocriticism, as Iovino argues, heed "the stories inbuilt in matter" (storied matter) (4). My focus on these theatrical performances' critical potential to creatively express the material world and the politics of human and nonhuman agencies on a variety of scales has been instrumental for this book. This framework gives voice to the unknown material forces and meanings unleashed by destabilising the binaries of science and politics, nature and culture, and fact and value, and by foregrounding the entanglement of collectives of human and nonhuman actants.

This research follows a trajectory from theatrical works that raise ecological and environmental awareness to those that, in a direct and obvious way, discard the humanist bias of theatre in favor of interwoven storylines of humans and nonhumans that help audiences connect affectively and ethically with overwhelming, abstract phenomena like anthropogenic climate change and mass extinction. For example, Chapters 2 and 3 chart a shift from traditionally anthropocentric theatrical forms to ecological ones. These chapters narrate representational changes across a range of plays, unveiling the relation of these changes to the extremely serious environmental issues of the Anthropocene by using ecocritical theories and concepts. Ecocritical interpretations reveal the ecological force of these plays. Chapters 4 and 5 show a similar relation, but more directly and pointedly underscore the uses of theatre as a vibrant medium for dealing with the ecological crisis in the contemporary world and imagining and experiencing the Anthropocene. Rather than merely focusing on how theatre engages with new representational modes in the context of the Anthropocene, I also consider how theatre fosters new practices of Anthropocene storytelling. Hence, through my analysis of these selected examples, I have outlined a new sketch of ecological theatre that demonstrates the importance of reflecting on how the plays and performances of the new millennium take up the storytelling challenges of the Anthropocene.

My study of the plays in Chapters 2 and 3 discloses theatre's ecological potential for interrogating common-sense notions of the human as a primary focus. Chapters 2 and 3 not only recalibrate these plays by applying an ecocritical lens to reveal their insights about the environment, but also reflect on how their materiality creates opportunities for increased ecocritical engagement and the decentring of human subjects. An ecocritical interpretation can emerge and take effect by proposing, to use Timothy Clark's term, a "de-humanizing" reading. Such a reading calls for "the drastic reconfigurations of context" and interpretation that "emerge with the concept of the Anthropocene" (Clark 117). In Chapter 2, I examine the ways in which Churchill's theatrical innovations and experiments seek to reconfigure conceptual bodies and spaces in the Anthropocene. By reading *The Skriker* and *Far Away* through a material ecocritical lens, we are able to give a voice to an ensemble of human and nonhuman forces and their interactions in the context of environmental change. This chapter shows how ecological theatre can dramatise nonhuman beings, things, and environmental forces, especially by repurposing figures from folklore and myth, as in *The Skriker*'s ecological allegory. The mutative nonhuman figures and, in particular, the shape-shifting and composite character of the Skriker, its frantic and fractured form of language play, and its circulation across time and space materialise a transformative and trans-corporeal environment, offering a theatrical model of hazardous landscapes and fragmented, toxic bodies intimately associated with ongoing changes in the material world. This chapter also demonstrates the way in which Churchill explores the dark underside of ecological interconnectedness in *Far Away*, where nonhuman species, things, matter, and natural forces definitely have agency, but often a violent and destructive agency that groups of humans use against each other. The vital materiality of the play's dark, composite environment evokes a reconceptualised material world in which the agency of humans is inextricably enmeshed with larger assemblages of human-nonhuman actants. The chapter concludes with a commentary on how a material ecocritical analysis of Churchill's works implies an innovative, non-anthropocentric kind of theatre, setting the stage for a more vibrant and collaborative account of the natural world at the dawn of the Anthropocene.

In Chapter 3, I pursue Clark's proposal for dehumanizing reading by analysing contemporary Australian theatre through the lens of ecocritical theory and producing new ways of representing Australian cultural and ecological engagement with the hyperobject of climate change as a distributed form of materialism. This chapter shows how Sewell's *It Just Stopped* and Bovell's *When the Rain Stops Falling* address the enormous scale and complexity of climate change through innovative theatrical narratives about human characters in extreme or absurd situations, in the first case, and of multiple generations of a family over time, in the second. Taking an ecocritical approach to these plays sheds light on the strangeness and destructiveness of daily activities when they are seen in relation to carbon emissions and, in fact, as part of the changing climate that is extended beyond our limited sense of time and space. I show how these plays' scripts, design, and staging can create a sense of deep time, pose challenges to our ways of thinking about the end of the World (Nature), and charge everyday objects like electrical appliances, raincoats, and umbrellas with apocalyptic resonance. While these plays have been omitted from discussions of ecological theatre, my book has (re)located these works within ecocritical theatre and revealed their reinscription of narratives based on identity, language, and dramatic form. These works are vital in showing that ecological awareness and interaction with the physical world are not purely marginal interests but are taken up by writers who are usually thought to be interested primarily in the social world, the political arena, and the family saga. These playwrights create a dark image of the human world estranged from the nonhuman environment to show the devastating cultural, psychological, and ecological consequences of such a distance.

Throughout the next two chapters, I discuss the ways in which theatre and performance practices can contribute to raising awareness of the climate and extinction crises as well as how they contribute to a paradigmatic shift towards a more contemplative approach in ecological theatre in the context of the Anthropocene. The ecodramaturgy of Lewis's *Song of Extinction* and Bilodeau's *Sila* in the fourth chapter further unpacks the ecological crises of extinction and climate change by eroding any familiar relation between cause and consequence, subject and object, human history and natural history, and intention and result. *Song of Extinction* dramatises a disordered cosmos – rather than a unified globe-shaped image – framed by a history of extinction as exemplified in the (eco)genocide scenarios in Bolivia and Cambodia, as well as a human being's loss of a loved one in Portland. These interconnected narratives provide alternatives to the current Anthropocene mass extinction in Earth's history, weaving a complex composition of extinction threads that are entangled in ongoing loops of intra-activity. To discuss extinction awareness in relation to climate change, I move to Bilodeau's *Sila*, in which nonhuman animals, human communities and individuals, culture, mythology, landscape, glaciers, and climatic patterns are progressively affected and dislocated. I show that the play's wide panoramas of multiple perspectives challenge the supposed limitation of theatrical space to personal and national scales. Instead, the play shifts the shape and force of ecological thinking towards what Clark has identified as derangements of scale. The collective challenge of climate change deranges scale, prompting awareness of a reconfiguration of space-time-matter and a process of intra-activity, to use Barad's words, in which an assemblage of human and nonhuman bodies come into play iteratively. The play involves agencies at diverse scales, from the human world to nonhumans speaking and acting as performative forces in ways that are central to the action.

Collective thinking and stories of multispecies intra-activity are expressed more directly and innovatively in the collaborative space of Hall's *A Brief History of Human Extinction* and Matreyek's *This World Made Itself* and *Infinitely Yours* in Chapter 5. The integration of science fiction into performance art in Hall's play might be considered a different ecocritical mode of narrating the enormity of the extinction crisis that yields new insights into the Anthropocene as a science fiction trope, "the idea of a planet terraformed by humans" (Heise 217). The play's immersion in the lives of the last members of the human species on the earth, the last living sea otter, and the world of fungi opens up a theatrical space of multispecies sympoiesis and collaboration, challenging bounded individualism and human exceptionalism and, to use Haraway's phrasing, generating an ethics of making kin as well as a practice of living and dying together on a damaged planet. Matreyek's live, staged, and embodied performances are situated at the cross-section of the cinematic and theatrical, fantastical and physical. Matreyek interacts with her kaleidoscopic moving images as a shadow silhouette, making scientific and biological ideas visible and material through her performing body. As a companion to *Infinitely Yours*, the semi-scientific and dream-like environment of *This World Made Itself* demonstrates the importance of deep time in the Anthropocene and narrates the intervention of past ecological forces in the present and future life of all entities. *Infinitely Yours* entails a dynamic performance of material meanings according to the changing environment and design conditions for multispecies flourishing in the face of the earth's climate risk. This performance practice invites us to think of ongoingness and staying with the troubled and muddled present, despite the irreversible environmental damage done in the past and the possibility of global mass extinction, including even that of humans. Corresponding with Haraway's idea of the Chthulucene, Matreyek's work enables the audience to visualise and experience how the human body makes kin with the changing ecosystem, different ways of sensing, living and dying, and becoming-with in n-dimensional time-spaces. This chapter makes a dramatic closure to this book, stressing the creative and significant role of theatre in enabling us to critically reflect on our place in the broader material world and on our way of living.

In spite of the significance of this book's conclusions, its focus is necessarily limited. Reflecting on the study of ecocritical theory in global Anglophone drama of the twenty-first century, however, might well bring to mind questions that this project has not contended with and that could act as catalysts for future scholarship and further research. The research presented in this book could be usefully extended by looking beyond the limits of English, Australian, and Canadian cultures. Indeed, examination of ecological plays from other cultures and geopolitical regions, in particular the Anglophone global South (Africa, Asia) and Indigenous communities, could yield new insights. In addition to the ways in which this study might inform the analysis of dramatic works that are critically attentive to the entanglement of the human and the nonhuman, whether or not they are understood as ecological works, this book lays the groundwork for innovative research in theatre studies with broader scope for ethnic and national diversity. Future research might expand the concept of the Anthropocene by considering some foundational difference (race, class, gender) in its definition and challenging the idea that the human species is a geological force that is detached from its relation to capitalism, differentials, and inequalities.

In the most general sense, each of the works of art discussed in this book is useful for anchoring the ecocritical theories to the Anthropocene in twenty-first century performance. This perspective offers a thoughtful and vital model for guiding present and future playwrights and eco-artists in their response to the urgency of the current anthropogenic environmental changes, such as climate change and the biodiversity crisis – to which the emergence of the Covid-19 pandemic is not unrelated.[1] Haraway underlines the urgency of thinking and living with(in) current troubling and turbid conditions:

> These times called the Anthropocene are times of multispecies, including human, urgency: of great mass death and extinction; of onrushing disasters, whose unpredictable specificities are foolishly taken as unknowability

> itself; of refusing to know and to cultivate the capacity of response-ability; of refusing to be present in and to onrushing catastrophe in time; of unprecedented looking away.
>
> (*Staying with the Trouble* 35)

The strength of each of these theatrical works' environmental commitment is evident in their evocation of the "thickness" of the present – that is, the importance of being immersed in the present – and highlights the necessity of cultivating the capacity of response-ability in a time of "unprecedented looking away." This book illuminates the role of theatre and performance in making sense of the challenges of the Anthropocene, offering a platform from which we can imagine the world's iterative performativity in which all bodies, human and nonhuman, are cast horizontally as agential forces on the theatrical world stage. At the end of this book, I offer my ecocritical readings of these plays and performances as a way of revealing an Anthropocenic turn in theatre and performance. I also hope to provoke further reflections on the flourishing of multispecies responsibilities and sensitivities in aesthetic and ethical terms.

Note

1 Seen in this light, for example, theatre might be able to deal productively with ecocritical narrations of the novel coronavirus pandemic, as the enormity of this worldwide infectious disease may be a consequence of human development in the Anthropocene and a result of anthropogenic forces that have changed the Earth's ecosystems (O'Callaghan-Gordo and Antó n. pag.).

References

Chakrabarty, Dipesh. "Postcolonial Studies and the Challenge of Climate Change." *New Literary History* 43.1 (2012): 1–18. Print.

Chaudhuri, Una. "The Fifth Wall: Climate Change Dramaturgy." Web blog post. *HowlRound.* HowlRound, 17 Apr. 2016. Web. 10 May 2016. <http://howlround.com/the-fifth- wall-climate-change-dramaturgy>.

Clark, Timothy. *Ecocriticism on the Edge: The Anthropocene as a Threshold Concept.* London: Bloomsbury Academic, 2015. Print.

Estok, Simon C. *The Ecophobia Hypothesis.* New York: Routledge, 2018. Print.

Haraway, Donna J. *Staying with the Trouble. Making Kin in the Chthulucene.* Durham, NC: Duke UP, 2016. Print.

Heise, Ursula K. *Imagining Extinction: The Cultural Meanings of Endangered Species.* Chicago: University of Chicago P, 2016.

Iovino, Serenella. *Ecocriticism and Italy: Ecology, Resistance, and Liberation.* London and New York: Bloomsbury Academic, 2016. Print.

O'Callaghan-Gordo, Cristina and Josep M. Antó. "COVID-19: The Disease of the Anthropocene." *U.S. National Library of Medicine.* National Center for Biotechnology Information, 15 May 2020. Web. 20 June 2020. <www.ncbi.nlm.nih.gov/pmc/articles/PMC7227607/>.

Read, Alan. *Theatre, Intimacy & Engagement: The Last Human Venue*. London: Palgrave Macmillan, 2009. Print.

Rose, Deborah Bird. "Introduction: Writing in the Anthropocene." *Australian Humanities Review* 49 (2009): 87. Web. 14 Apr. 2013. <http://australianhumanitiesreview.org/2009/11/01/introduction-writing-in-the-anthropocene/>.

Index

Note: Numbers in *italics* indicate a figure.